# Essays in Biochemistry

# Essays in Biochemistry

Edited for The Biochemical Society by

## R. D. Marshall

Division of Biochemistry
Department of Bioscience and Biotechnology
University of Strathclyde
Glasgow
Scotland

## K. F. Tipton

Department of Biochemistry
Trinity College
University of Dublin
Dublin 2
Eire

## Volume 23

**1987**

Published for The Biochemical Society by Academic Press
London, San Diego, New York, Boston
Sydney, Tokyo, Toronto

ACADEMIC PRESS LIMITED
24/28 Oval Road
London NW1 7DX

U.S. Edition published by
ACADEMIC PRESS INC.
San Diego
CA 92101

ISBN 0-12-158123-3
ISSN 0071-1365

Photoset by Paston Press, Loddon, Norfolk
Printed in Great Britain by
Galliard Printers,
Great Yarmouth

# Biography

*Hugh Nimmo* graduated from the University of Cambridge in 1970 and remained there for a further three years to do a PhD in Biochemistry under the supervision of Dr Keith Tipton. In 1973 he moved to Dundee University where he was a member of Dr Philip Cohen's group and was introduced to the delights of protein phosphorylation. Since 1977 he has been on the staff of the Biochemistry Department at Glasgow University. His research interest concerns the mechanisms by which enzyme activities are regulated and the roles played by these mechanisms in controlling metabolic pathways. He has been involved in studies of protein phosphorylation in mammalian, plant and prokaryotic systems.

*Michael G. Palfreyman* graduated in Pharmacy with an Honours degree from the University of Nottingham, UK and then studied for his doctorate at the same university department under the supervision of Professors J. C. Crossland and B. E. Leonard. The thesis work was on neurochemical changes during convulsive activity. On completion of his doctorate he went to Beecham Pharmaceuticals Research Division where he became Head of the Biochemical Pharmacology Unit. In 1976 he joined the Centre de Recherche Merrell International in Strasbourg, France as an Established Investigator and in 1983 became the Head of the Neurochemistry and Behavioral Section. In 1986 he moved to the Merrell Dow Research Institute, Cincinnati Center in the USA, as Head of CNS Research and in 1987 became Director of Pharmacological Sciences.

Throughout his scientific career he has been interested in enzyme inhibitors and has published extensively on many aspects of inhibitor design and use.

*Philippe Bey* graduated from the École Nationale Superieure de Chimie de Strasbourg, France in 1963. He then joined the Centre National de la Recherche Scientifique and worked with Professor G. Ourisson at Louis Pasteur University in Strasbourg on the mechanisms of the enone-benzene rearrangement and received his PhD in 1967. On completion of his doctorate, he served for two years as Assistant Professor of Chemistry at the Lebanese University at Beirut, before spending two years with Professor R. E. Ireland at the California Institute of Technology where he worked on the total synthesis of pentacyclic triterpenes via biomimetic cyclization of polyenes. In 1971, he joined the Centre de Recherche Merrell International in Strasbourg, France as a Senior Scientist and in 1979 became the Head of the Chemistry Department. He moved to the Merrell Dow Research Institute, Cincinnati Center in 1984 where he is presently the Director of Chemical Sciences.

One of his main research activities has been the design and synthesis of specific inhibitors for enzymes of thereapeutic relevance. Throughout his scientific career he has been interested in biologically active organofluorine compounds.

*Albert Sjoerdsma* did his academic training at the University of Chicago, where he received four degrees; a PhB, a BS in Physiology, an MD and a PhD in Pharmacology. He is a Diplomate of the American Board of Internal Medicine. He was in the US Public Health Service with the National Institutes of Health in Bethesda, Maryland from 1951–1971, where he served as Chief of the Experimental Therapeutics Branch of the National Heart and Lung Institute. In 1971 he accepted the position of Vice President and Director of Research at the Centre de Recherche Merrell International in Strasbourg, France. He is currently President of the Merrell Dow Research Institute, which is composed of research centers in Egham, England; Gerenzano, Italy; Strasbourg, France; and Indianapolis and Cincinnati in the USA.

In the field of clinical pharmacology he has achieved widespread recognition for his contributions to experimental medicine and the development of new thereapeutic agents. His current interests have personally been focused on inhibitors of polyamine biosynthesis as therapy for cancer and lethal protozoal infections such as African sleeping sickness and *Pneumocystis carinii* pneumonia in AIDS. He is an editor of a book on this topic published in 1987 by Academic Press – *Inhibition of Polyamine Metabolism: Biological Significance and Basis for New Therapies.*

*David Morgan* began his research career in the Surgical Unit at St Mary's Hospital Medical School, London, where he took part, with Dr K. J. Kingsbury, in a study on lipid metabolism in relation to atheroma. A transfer to the Department of Chemical Pathology led to work on amino sugar chemistry with Professor A. Neuberger. He became interested in polyamines after moving to the Clinical Research Centre (at Harrow, Middlesex) to work with Dr A. C. Allison on the effects of activation on macrophage enzymes. The discovery that these cells contain a polyamine oxidase, and similarities between this enzyme and the amine oxidase that occurs in human serum in pregnancy, initiated a joint investigation with Professor G. Illei, a visiting worker from Hungary, of some of the properties and functions of this enzyme. His main interests are polyamine oxidase and the effects of its products on cell growth and proliferation. He is currently studying the biochemical responses of human vascular endothelial cells to various stimuli (including polyamines) that may affect their growth and function. He is a Founder of the British Polyamine Group.

*C. B. Pickett* is currently Director of the Department of Molecular Pharmacology and Biochemistry at Merck Sharp & Dohme Research Laboratories in Rahway, New Jersey. He received his BSc degree in Biology from California State University, and his PhD degree in Cell Biology from the University of California, Los Angeles. His research interests are in the field of regulation of gene expression, focusing on genes encoding drug metabolizing enzymes. He has served as a member of the National Institutes of Health's Physical Biochemistry Study Section and is on the Editorial Board of the Journal of Biological Chemistry.

# Conventions

The abbreviations, conventions and symbols used in these Essays are those specified by the Editorial Board of *The Biochemical Journal* in *Policy of the Journal and Instructions to Authors* (see first issue in latest calendar year). The following abbreviations of compounds, etc., are allowed without definition in the text.

ADP, CDP, GDP, IDP, UDP, XDP, dTDP: 5'-pyrophosphates of adenosine, cytidine, guanosine, inosine, uridine, xanthosine and thymidine.

AMP, etc.: adenosine 5'-phosphate, etc.

ATP, etc.: adenosine 5'-triphosphate, etc.

CM-cellulose: carboxymethylcellulose

CoA and acyl-CoA: coenzyme A and its acyl derivatives

Cyclic AMP, etc.: adenosine 3',5'-cyclic phosphate, etc.

DEAE-cellulose: diethylaminoethylcellulose

DNA: deoxyribonucleic acid

Dnp-: 2,4-dinitrophenyl-

Dns-: 5-dimethylaminonaphthalene-1-sulphonyl-

EDTA: ethylenediaminetetra-acetate

FAD: flavin adenine dinucleotide

FMN: flavin mononucleotide

GSH, GSSG: glutathione, reduced and oxidized

NAD: nicotinamide adenine dinucleotide

NADP: nicotinamide adenine dinucleotide phosphate

NMN: nicotinamide mononucleotide

$P_i$, $PP_i$: orthophosphate, pyrophosphate

RNA: ribonucleic acid (see overleaf)

TEAE-cellulose: triethylammonioethylcellulose

tris: 2-amino-2-hydroxymethylpropane-1,3-diol

The combination $NAD^+$, NADH is preferred.

The following abbreviations for amino acids and sugars, for use only in presenting sequences and in Tables and Figures, are also allowed without definition.

## Amino acids

| | | |
|---|---|---|
| Ala: alanine | Asx: aspartic acid or | Cys: cystine or cysteine |
| Arg: arginine | asparagine | |
| | (undefined) | |
| Asn: asparagine | | Gln: glutamine |
| Asp: aspartic acid | | Glu: glutamic acid |

Glx: glutamic acid     Ile: isoleucine     Pro: proline
    or glutamine     Leu: leucine     Ser: serine
    (undefined)
Gly: glycine     Lys: lysine     Thr: threonine
His: histidine     Met: methionine     Trp: tryptophan
Hyl: hydroxylysine     Orn: ornithine     Tyr: tyrosine
Hyp: hydroxyproline     Phe: phenylalanine     Val: valine

*Sugars*

Ara: arabinose     Glc*: glucose
dRib: 2-deoxyribose     Man: mannose
Fru: fructose     Rib: ribose
Fuc: fucose     Xyl: xylose
Gal: galactose

*Where unambiguous, G may be used.

Abbreviations for nucleic acids used in these essays are:

mRNA: messenger RNA
nRNA: nuclear RNA
rRNA: ribosomal RNA
tRNA: transfer RNA

Other abbreviations are given on the first page of the text, or at first mention.

References are given in the form used in *The Biochemical Journal*, the last as well as the first page of each article being cited, and, in addition, the title. Titles of journals are abbreviated in accordance with the system employed in the *Chemical Abstracts Service Source Index* (1979) and its Quarterly Supplement (American Chemical Society).

## Enzyme Nomenclature

At the first mention of each enzyme in each Essay there is given, whenever possible, the number assigned to it in *Enzyme Nomenclature: Recommendations (1984) of the Nomenclature Committee of the International Union of Biochemistry on the Nomenclature and Classification of Enzyme-catalysed Reactions*, published for the International Union of Biochemistry by Academic Press, New York and London, 1979. Enzyme numbers are given in the form EC 1.2.3.4. The names used by authors of the Essays are not necessarily those recommended by the International Union of Biochemistry.

# Contents

# Preface

Many areas of biochemistry are advancing so rapidly that it is easy for major developments to occur over the course of a decade or so without the non-specialist or writers of textbooks being aware of what is happening. The Essays in this volume demonstrate this feature.

Protein phosphorylation has been recognized for a number of years as forming part of one of the mechanisms for regulating metabolism in animals. For a long time there was considerable dubiety as to whether control of metabolism of this type occurs naturally in bacteria, and knowledge that indeed it does began to emerge only about 10 to 15 years ago. H. G. Nimmo in his Essay gives an up-to-date account of the controls exerted on bacterial metabolism through protein phosphorylation. The topic has clear implications for the use of prokaryotic cells in processes designed for commercial use.

Attempts to understand the ways in which xenobiotics act have been pursued for a number of years, for a variety of reasons, not least of which are those concerned with the environment, both natural and man-made. The powerful, toxic effects of the South African plant species, *Dichapetalum*, which has led to the deaths of many cattle ingesting it, have been known for decades. About 40 years ago, the toxicity was shown to result from enzymic conversion in the animal of the fluoroacetate, which the plant contains, to fluorocitrate, an inhibitor of a different enzyme (aconitase) in the citric acid cycle. But more direct, and more specific, enzyme-produced inhibitors have been examined over the past 10 to 15 years. These inhibitors are substances produced by the action of an enzyme on a foreign substrate, and the product formed inhibits the action of that *same* enzyme on its natural substrate. M. G. Palfreyman, P. Bey and A. Sjoerdsma discuss this phenomenon, and show the value of studies of this nature both as a research tool and for the development of new therapeutic agents.

Although the polyamines appear to be important for eukaryotic cell proliferation and growth, a subject of major importance, these substances are usually given but scant attention in textbooks of biochemistry, including those of cell and molecular biology. If full use is to be made of animal cell cultures for the production of valuable substances in the service of Man, more attention must be paid to the polyamines. The Essay by D. M. L. Morgan addresses some metabolic aspects of these substances, and draws attention to the likelihood that *each* of them may have a *number* of different functions.

Systematic studies by R. Tecwyn Williams over the decades following the Second World War led to the general ideas concerning the mechanisms whereby many foreign compounds are eliminated from the body. Interspecies differences were studied and in particular the nature of the conjugation step. But it is only relatively recently that the complexity of this step in both cellular and genetic terms has begun to emerge. C. B. Pickett in his Essay writes about the enzymes for one type of conjugation, in particular the glutathione S-transferase genes. The topic is at a stage of exciting development.

The authors give of their time and energy to write in what we believe to be an informative and attractive manner, providing the main trends in their specialist fields. Constructive criticisms, and even congratulatory comments for the authors, are always welcome, as are suggestions for further topics. Please write and let us know your views.

R. D. Marshall & K. F. Tipton
*November 1987*

# Regulation of Bacterial Metabolism by Protein Phosphorylation

## HUGH G. NIMMO

*Department of Biochemistry, University of Glasgow, Glasgow G12 8QQ, UK*

## I. Introduction

Protein phosphorylation has long been recognized as of major importance in the control of many facets of the behaviour of eukaryotic cells (e.g. refs 1 and 2). In contrast, over a number of years, there was considerable doubt that protein phosphorylation occurred in prokaryotes at all. For example, the kinase activity detected in extracts of *Escherichia coli*[3] was later found to be a polyphosphate kinase, not a protein kinase.[4] The protein kinase activity detected in cells infected with bacteriophage T7 was shown to be a

ESSAYS IN BIOCHEMISTRY Vol. 23
ISBN 0 12 158123-3

phage gene product.[5,6] Thus it became widely assumed that prokaryotic cells normally contained no protein kinase activity (e.g. ref. 7).

However, in 1978, Wang and Koshland[8] showed this to be incorrect when they demonstrated the existence of at least four $^{32}$P-containing, ribonuclease-resistant but pronase-sensitive species after pulse-labelling of *Salmonella typhimurium* cells with $^{32}$P$_i$. These proteins could be radioactively labelled *in vitro* by the addition of [$\gamma$-$^{32}$P]ATP to crude cell extracts, but not by addition of either [$\alpha$-$^{32}$P]ATP or [$^{14}$C]ATP; the $^{32}$P was incorporated so as to form phosphoserine and phosphothreonine residues.[8] Manai and Cozzone[9] obtained similar results with *E. coli* shortly afterwards.

Recent work has revealed that the phenomenon of protein phosphorylation is quite common in prokaryotes; for example, in the system best studied (that of *E. coli*) over 100 phosphorylated proteins have been resolved.[10] This is comparable to the number of phosphorylated proteins detected in the soluble fraction of rat hepatocyte extracts.[11] However, very few of the phosphorylatable proteins in prokaryotes have yet been identified. This review briefly discusses studies of protein phosphorylation in intact prokaryotic cells and describes the metabolic roles of those systems in which the target of phosphorylation has been identified.

## II.  The Occurrence of Protein Phosphorylation in Prokaryotes and the Existence of Distinct Protein Kinases

### A.  $^{32}$P-LABELLING STUDIES IN INTACT CELLS

Several groups have carried out broadly similar experiments with different organisms in which cultures are labelled with $^{32}$P$_i$ in a low-phosphate medium. Cells are harvested and broken and the proteins are separated by gel electrophoresis. Proteins that contain $^{32}$P-phosphate are revealed by autoradiography.

In these experiments it is clearly essential to assess the nature of the protein-bound, $^{32}$P-containing group. Protein kinases catalyse the formation of phosphomonoester derivatives of serine, threonine or tyrosine residues. Phosphate groups can become covalently attached to proteins, usually on histidine, aspartic acid or glutamic acid residues, during the course of catalytic turnover. It is also possible that phosphate groups could become covalently bound to proteins as constituents of ADP-ribosyl, adenyl or uridyl groups. Various criteria have been used to identify those proteins that contain $^{32}$P-phosphate as a phosphomonoester. These include resistance to neutral hydroxylamine, hot trichloracetic acid and phosphodiesterase, and sensitivity to acid or alkaline phosphatase with direct identification of a phosphoaminoacid after partial acid hydrolysis.

A systematic survey of protein phosphorylation in *E. coli* has recently been presented by Cortay *et al.*[10] Proteins were separated by two-dimensional gel electrophoresis. Cells growing exponentially on glucose or glycerol gave similar patterns of phosphorylated proteins. In cultures at stationary phase after growth on glucose or glycerol, a larger number of phosphorylated proteins was detectable, and the intensities of some of the spots were greater than during exponential growth. (The latter observation cannot, of course, be interpreted as an increased stoichiometry of phosphorylation.) Heat shock, addition of ethanol or starvation of amino acids reduced the number of phosphorylated proteins that were detectable, but did not cause the appearance of any previously undetected species.

The largest number of phosphorylated proteins was observed following addition of acetate for 90 min to cells grown to stationary phase on glycerol. In this case a total of 128 radioactive spots could be detected. This is considerably greater than the number of phosphorylated proteins reported in earlier studies of *E. coli*,[9,12,13] most probably because of improvements in the resolution and detection of phosphorylated proteins. Several factors could, of course, cause a single protein to give rise to multiple, $^{32}$P-containing spots on a two-dimensional gel. However, it is also likely that several phosphorylatable proteins are present in such low amounts in cells that they escape detection. One conclusion drawn from this work must be that protein phosphorylation is more widespread in *E. coli* than had previously been suspected.

Cortay *et al.*[10] identified the phosphoaminoacids present in 20 of the phosphorylated *E. coli* proteins. All except one contained phosphoserine and, of these 19, three also contained phosphothreonine. As had been reported earlier,[14] one protein contained only phosphotyrosine. This observation is of great interest in view of the attention currently being devoted to tyrosine-specific phosphorylation in mammalian systems (e.g. ref. 2). This is the only known example of phosphorylation of a bacterial protein on a tyrosine residue.[10,14] The phosphotyrosine-containing protein was present in a crude preparation of ribosomes but its electrophoretic mobility did not match that of any known ribosomal protein. It was also present in a crude membrane pellet, but it was lost from the membranes when they were further purified by sucrose density gradient centrifugation.[10] Cortay *et al.*[10] suggested that this protein might be located at the interface between membranes and ribosomes but its identity is unknown.

Studies of the subcellular location of the other phosphorylated proteins revealed that the great majority were soluble.[10] Three of the proteins were firmly bound to membranes but none were associated with nucleoids. Earlier reports that several ribosomal proteins[9] and subunits of RNA polymerase[13] could be phosphorylated were not confirmed.

Cortay *et al.*[10] could identify only two of the many phosphorylated proteins that they observed; these were isocitrate dehydrogenase (ICDH) and the product of the *dnaK* gene (see Sections III and VI respectively). The spot corresponding to ICDH was one of 22 "very intense" spots detected after the addition of acetate to cells grown on glycerol. It is not known whether ICDH was fully phosphorylated in this experiment. However, results with another strain of *E. coli* suggest that under these conditions ICDH represents perhaps 0·5–1·0% of the soluble protein of the cell[15] and that it may be phosphorylated to about 0·75 molecules/subunit.[16] The detection of several other spots of comparable intensity therefore argues that some other quite abundant *E. coli* proteins can be phosphorylated or that some proteins can be very heavily phosphorylated (or both).

Protein phosphorylation has also been demonstrated in *E. coli* in other growth conditions, and in various other bacterial species. Examples include *E. coli* during anaerobic growth on glucose with nitrate as electron acceptor,[17] *Salmonella typhimurium* growing on glucose or acetate,[18] *Clostridium sphenoides* growing anaerobically on citrate[19] and *Rhodocyclus gelatinosa* during anaerobic growth on citrate in the light.[20] In each case, fewer phosphorylated proteins (8–15) were resolved and detected than in the work on *E. coli* discussed above. In several cases the number, identity and intensity of the phosphorylated proteins present in the cells were found to depend on the carbon source or growth conditions.[13,18,20] These changes could, of course, be brought about either by changes in the phosphorylation state of the proteins involved or by changes in the intracellular concentrations of the proteins.

The major limitation on studies of protein phosphorylation in intact cells is that so few of the phosphorylated proteins have been identified. Studies on the proteins that have been identified are discussed below (Sections III–V).

## B. STUDIES OF BACTERIAL PROTEIN KINASES

Following the demonstration of protein phosphorylation in intact bacterial cells, two groups attempted to fractionate bacterial protein kinases without prior knowledge of the identities of their protein substrates. (Studies of the kinases whose substrates have been identified are described in Sections III and IV.) Wang and Koshland[21] first labelled proteins in extracts of *Salmonella typhimurium* by addition of [$\gamma$-$^{32}$P]ATP. They fractionated these extracts to discover the chromatographic behaviour of the substrates of the phosphorylation reactions. They then fractionated unlabelled extracts and incubated the resulting fractions, alone or in combinations, with [$\gamma$-$^{32}$P]ATP. In several cases they were able to separate protein kinases from their substrates. The kinases could also be distinguished by virtue of difference in sensitivity to inhibitors such as PP$_i$, AMP and GTP. These experiments revealed the existence of at least four different protein kinases and two different protein phosphatases. These protein kinases were able to phosphorylate *in vitro* some, but not all, of the proteins that could be labelled in intact cells.[21] They were unable to phosphorylate histones, casein and phosvitin,[8,21] which are frequently used as substrates for eukaryotic protein kinases. This largely explains the early failure to detect the existence of protein kinases in prokaryotes.

Enami and Ishihama[13] observed that, among the proteins that can be phosphorylated in intact *E. coli* cells, there is one of subunit $M_r$ 90 000. They were able to resolve this protein from the protein kinase responsible for its phosphorylation by ion-exchange chromatography. They purified the kinase further and obtained an apparently homogeneous protein of native $M_r$ 120 000 that contained two subunits of $M_r$ 66 000 and 61 000. This protein was able to transfer the $\gamma$-phosphate of ATP to the partially purified $M_r$-90 000 protein *in vitro*.[13] The identity of the substrate and the stoichiometry, role and regulation of this phosphorylation are still not known.

### III. Phosphorylation of Isocitrate Dehydrogenase and Control of Flux through the Tricarboxylic Acid Cycle and the Glyoxylate Bypass

#### A. INTRODUCTION

Growth of microorganisms on acetate requires the two enzymes of the glyoxylate bypass, isocitrate lyase and malate synthase, to generate the intermediates of the tricarboxylic acid (TCA) cycle that are used as precursors for biosynthesis[22] (Fig. 1). In *E. coli*, isocitrate lyase and malate synthase A are induced during growth on acetate or fatty acids (a distinct isoenzyme, malate synthase B, is induced during growth on glycollate) and, under these conditions, there is therefore competition for the available isocitrate between ICDH and isocitrate lyase. The kinetic properties of these two enzymes do not indicate how the competition might be resolved. *E. coli* contains only an NADP-linked ICDH that is not subject to allosteric control, and while an early report suggested that phosphoenolpyruvate (PEP) was a feedback inhibitor of isocitrate lyase,[23] consideration of the intracellular concentrations of isocitrate and phosphoenolpyruvate indicated that this effect is not physiologically significant.[24]

The first indication that ICDH could be regulated by reversible inactivation came from the work of Bennett and Holms[10,25] on the levels of TCA cycle enzymes in *E. coli* ML308 during growth on different carbon sources. After cessation of growth on limiting glycerol, the specific activities of ICDH, malate dehydrogenase and 2-oxoglutarate dehydrogenase in crude cell extracts remained constant for several hours into stationary phase. After cessation of growth on limiting glucose, however, the specific activities of

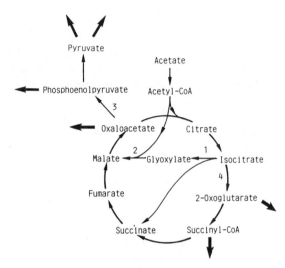

**Fig. 1.** The TCA cycle and the glyoxylate bypass. The enzymes are as follows: (1) isocitrate lyase; (2) malate synthase; (3) phosphoenolpyruvate carboxykinase; (4) ICDH. Heavy arrows represent fluxes to biosynthesis.

malate dehydrogenase and 2-oxoglutarate dehydrogenase remained stable, whereas that of ICDH fell to 20% of its original value over 2 h and then recovered to 75% over the next 2 h.[25] This occurred because *E. coli* excretes acetate during growth on glucose but not on glycerol. After growth on glucose ceased, the enzymes of the glyoxylate bypass were induced, and the acetate was then oxidized. The specific activity of the ICDH declined during the period of adaptation to and use of acetate, but rose again after exhaustion of the acetate.[25]

The specific activity of ICDH in cells growing exponentially on acetate was only some 30% of the value for growth on glucose or glycerol, but it could be increased rapidly in various ways. For example, the addition of 1 mM pyruvate to cells growing on acetate gave a rapid four-fold activation of ICDH that persisted until the supply of pyruvate was exhausted, whereupon the ICDH activity reverted to the value characteristic of growth on acetate. This effect was independent of protein synthesis and was not caused by low molecular weight effectors on ICDH.[16] Addition of some other carbon sources to cells growing on acetate could also cause reactivation of ICDH. Bennett and Holms[16] concluded that a reversible covalent modification of ICDH might be involved.

Phosphorylation was first implicated by the work of Garnak and Reeves.[26,27] They added $^{32}P_i$ to a culture of *E. coli* K12 grown to stationary phase on limiting glycerol in a low-phosphate medium. Addition of acetate to the culture caused both a partial reduction in ICDH activity and incorporation of $^{32}$P-phosphate (as phosphoserine) into ICDH protein. At this stage, however, the relationship between the phosphorylation and the activity of ICDH was not established, and isolation of the protein kinase and protein phosphatase involved was required.

## B. COMPONENTS OF THE SYSTEM

LaPorte and Koshland[28] made an enormous contribution when they showed that the activities responsible for catalysing the phosphorylation and dephosphorylation of ICDH co-purify. ICDH kinase/phosphatase is a dimer of $M_r$ 66 000 subunits. It appears to be the only enzyme in *E. coli* that can either phosphorylate or dephosphorylate ICDH.[28,29]

The fact that ICDH kinase/phosphatase contains a single type of bifunctional polypeptide chain was established at the DNA level[30] rather than at the protein level, largely because of the low abundance of the enzyme. The genes encoding isocitrate lyase (*aceA*) and malate synthase A (*aceB*) are part of the glyoxylate bypass operon located at 90 min on the *E. coli* chromosomal map.[31,32] Genetic studies showed that the gene encoding the kinase/phosphatase (*aceK*) was part of this operon, the gene order being *aceBAK*.[33] The operon has now been cloned from two strains of *E. coli*, K12 (W. D. Nunn, cited in ref. 30) and ML308.[34] The *aceK* gene was subcloned and deletion analysis showed that its coding region comprised some 1800 base pairs.[30] This region can only code for a single polypeptide of $M_r$ 66 000, which therefore bear both the kinase and the phosphatase activities. However, a major question concerning this enzyme, namely whether it contains two distinct active sites or a single site that can catalyse both reactions, has yet to be answered.

ICDH kinase transfers the $\gamma$-phosphoryl group of ATP to a serine residue of ICDH with a stoichiometry of one per subunit.[29,35] This results in essentially complete loss of ICDH activity, measured under $V_{max}$ conditions. The sequence round the phosphorylation site has been established;[36,37] it bears no relation to the sequences round sites phosphorylated by eukaryotic cyclic AMP-dependent protein kinases. Recent studies with intact cells[38] suggest that ICDH can also be phosphorylated at a threonine residue with a low stoichiometry, but it is not known whether this affects the activity of the enzyme.

ICDH phosphatase requires a divalent metal ion and either ADP or ATP for activity.[28,29] It catalyses the release of $P_i$ from phosphorylated ICDH and, provided that the kinase is not active under the conditions of the experiment, can reactivate the dehydrogenase fully.[29] The adenine nucleotide requirement of the phosphatase was surprising, and led to the speculations[28] that the active sites of the kinase and phosphatase are distinct and that a phosphatase domain arose through duplication of a primordial kinase gene followed by a gene fusion. This theory will be testable as soon as the nucleotide sequence of the *aceK* gene is known.

The molecular mechanism by which phosphorylation affects the activity of ICDH seems unusual. Many of the eukaryotic enzymes regulated by phosphorylation are allosteric proteins, and phosphorylation typically affects affinities for ligands rather than $V_{max}$ values. In contrast, phosphorylated ICDH is almost completely inactive, and fluorescence titration experiments showed that it cannot bind NADPH.[39] Its conformation is, however, similar to that of active enzyme containing bound coenzyme.[39] It has been suggested that phosphorylation occurs close to or at the coenzyme-binding site. Interaction between the negative charge of the phosphate group and a positive charge at the coenzyme-binding site may elicit a conformational change similar to that induced by the binding of coenzyme.[39,40] Determination of the three-dimensional structure of ICDH will be required to test this hypothesis fully.

## C. FACTORS THAT AFFECT THE PHOSPHORYLATION OF ISOCITRATE DEHYDROGENASE

ICDH kinase and ICDH phosphatase can be active simultaneously both *in vitro*[29] and in intact cells.[41,42] The phosphorylation state and activity of ICDH therefore represent the steady-state balance achieved by the kinase and phosphatase. This emphasizes the potential importance of effectors of these activities.

Several metabolites both inhibit the kinase and activate the phosphatase, including ADP, AMP, isocitrate, oxaloacetate, 2-oxoglutarate, phosphoenolpyruvate, 3-phosphoglycerate and pyruvate.[35,43] Each compound is capable of giving essentially complete inhibition of the kinase and two- to three-fold activation of the phosphatase at saturation.[43] These effects are probably mediated by binding of the effectors to the kinase/phosphatase but the number of distinct regulator-binding sites is not known.[43] $NADP^+$ and NADPH both inhibit ICDH kinase. If the coenzyme binding site is indeed close to the phosphorylation site, one possible explanation of this inhibition is that binding of NADP(H) to ICDH prevents phosphorylation of the enzyme by the kinase. In addition, NADPH is a strong inhibitor of the phosphatase.[43] These results are summarized in Fig. 2.

Since several metabolites affect the activities of the kinase and phosphatase in opposite directions, the phosphorylation state of ICDH should respond very sensitively to small changes in the intracellular concentrations of these compounds. Another factor that may affect the sensitivity of the system involves the phenomenon of "zero-order ultrasensitivity", first discussed by Goldbeter and Koshland.[44] If either or both of the converter enzymes in a covalent modification mechanism are close to saturation with

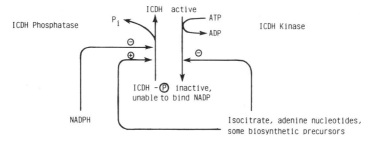

**Fig. 2.** Control of the phosphorylation state of isocitrate dehydrogenase. ⊕ and ⊖ represent stimulatory and inhibitory effects respectively. The precursors that affect the system are listed in the text.

their protein substrate (i.e. in the zero-order range), the system can respond (in terms of the degree of modification of the protein substrate) much more sensitively than if both the converter enzymes are far from saturation (i.e. in the first-order range). It has been shown that zero-order ultrasensitivity can occur in the ICDH system *in vitro*,[35] and the intracellular concentration of ICDH is sufficiently high that the phenomenon may also be important in intact cells.

## D. ROLE AND REGULATION OF THE PHOSPHORYLATION SYSTEM IN INTACT CELLS

Holms and Bennett[16,25] proposed that the role of the partial inactivation of ICDH during growth on acetate was to facilitate flux through the glyoxylate bypass, and this hypothesis has been widely accepted. ICDH can be phosphorylated even in mutants lacking a functional glyoxylate bypass,[45] but this observation is compatible with the hypothesis above because the kinase/phosphatase responds to a wide variety of metabolites and not merely to those unique to the glyoxylate bypass.

The significance of the phosphorylation of ICDH in intact cells has been carefully studied during the reversible activation of ICDH caused by the addition of pyruvate to *E. coli* growing on acetate.[41] It was shown, using cells prelabelled with $^{32}P_i$, that the phosphorylation state and the activity of ICDH were inversely related and that the reversible activation and deactivation of the enzyme resulted entirely from dephosphorylation and rephosphorylation. Moreover, the only serine residue phosphorylated in the enzyme in intact cells was identical with the residue labelled by purified ICDH kinase/phosphatase *in vitro*.[41] The importance of this work lies in showing that the changes in ICDH activity can safely be interpreted in terms of changes in the activity of the kinase/phosphatase, presumably mediated by changes in the intracellular concentration of one (or more) of its effectors.

Two groups independently proposed similar mechanisms that explain how phosphorylation of ICDH affects the division of flux between the TCA cycle and the glyoxylate bypass.[40,43,46,48] The key points are (1) that the phosphorylation and inactivation of ICDH during growth on acetate renders it rate-limiting in the TCA cycle, (2) that this reduces the rate of flux through the cycle and causes an increase in the intracellular concentration of isocitrate and (3) that this in turn results in an increase in flux through isocitrate lyase.

Several lines of evidence, obtained from both experimental and theoretical studies, support these views. A comparison of the $V_{max}$ of ICDH (estimated from *in vitro* assays) with the measured flux through the enzyme strongly suggests that ICDH is rate-limiting in the TCA cycle in cells growing on acetate.[48] The intracellular content of isocitrate is much higher during growth on acetate than on glucose or glycerol.[24,49,50] The $K_m$ of isocitrate lyase for isocitrate is much higher than that of ICDH and it is high relative to the intracellular content of isocitrate.[47,49–51]

LaPorte *et al.*[47] carried out a theoretical analysis of the fluxes through the two limbs of a branch point as a function of the kinetic parameters of the two competing enzymes. They showed that the partitioning of flux was extremely sensitive to changes in the activity of one of the enzymes if its $K_m$ for the common substrate was much lower than that of the competing enzyme. These conditions apply to ICDH and isocitrate lyase and this "branch point effect" is yet another factor that may increase the sensitivity of the control mechanism in intact cells.[47]

Walsh and Koshland[46] developed techniques that allowed them to determine the net rates of carbon flux through the central metabolic pathways during growth on acetate. They then studied the transition caused by the addition of glucose to acetate-grown cells.[48] This perturbation caused dramatic decreases in the flux through isocitrate lyase (by a factor of 150) and the intracellular content of isocitrate. These changes apparently resulted from a 4-fold increase in ICDH activity (caused by dephosphorylation) and a 5·5-fold decrease in the flux through citrate synthase.

The metabolic signals that trigger changes in the phosphorylation state of ICDH have been analysed carefully in the case of the reversible activation caused by addition of pyruvate to acetate-grown cells.[50,52] The importance of the intracellular isocitrate concentration (see above) and the observation that isocitrate is a powerful effector of the kinase/phosphatase led to the suggestion that isocitrate might be one signal.[40,43] Addition of pyruvate to acetate-grown cells is indeed accompanied by a transient two-fold increase in the intracellular content of isocitrate;[50] the rates of the changes in isocitrate were compatible with the view that these changes were partially or fully responsible for the changes in the phosphorylation state of ICDH. Pyruvate itself is an effector of the kinase/phosphatase *in vitro*.[43] Studies of the addition of pyruvate to acetate-grown mutants of *E. coli* lacking various enzymes of pyruvate metabolism led to the conclusion that pyruvate itself must be a physiologically important signal.[52]

Thus the role of the phosphorylation system seems to be to maintain the intracellular concentration of isocitrate at a level than can sustain the necessary flux through isocitrate lyase to biosynthetic precursors. At least two of the effectors of the kinase/phosphatase seem to be physiologically significant. The roles of other effectors have also been considered.[53] Among the metabolites that activate ICDH phosphatase and inhibit ICDH kinase are several biosynthetic precursors. High levels of these compounds should increase ICDH activity and thus reduce the flux through the glyoxylate bypass. These effects can be seen as a feedback mechanism to control the flux through the bypass, which is responsible for the generation of the precursors. The effects of ADP and AMP can be regarded as a mechanism to increase flux through the TCA cycle (and hence energy generation) when the cell's energy change is low.

Biosynthesis of course requires ATP and reducing power, so the division of flux between the TCA cycle (provision of energy and reducing power) and the glyoxylate bypass (supply of precursors) must be controlled precisely. Figure 3 is a heavily-oversimplified diagram that summarizes some of the interrelationships involved in the control of ICDH. It is likely that the relative importance of the different signals that can affect the phosphorylation state of ICDH varies under different conditions. However, the point to be stressed is that the system appears able to integrate a wide variety of metabolic information, in the shape of metabolite concentrations, and to adjust the phosphorylation state of ICDH to optimize the division of flux between the TCA cycle and the glyoxylate bypass.

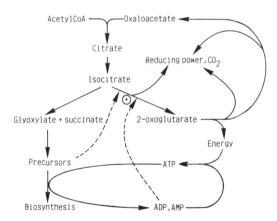

**Fig. 3.** Overview of the control of flux between the TCA cycle and the glyoxylate bypass. ⊕ represents stimulation.

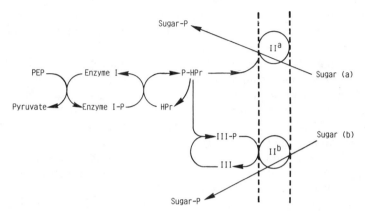

**Fig. 4.** Generalized view of the phosphotransferase system. P-HPr represents HPr phosphory-lated on a histidine residue. In some cases (a) this transfers its phosphoryl group directly to the sugar-specific, membrane-bound Enzyme II (II). In other cases (b) transfer is to a third soluble component, Enzyme III (III).

## IV. Phosphorylation of HPr and Regulation of the Phosphotransferase System in Streptococci

In most anaerobic and facultatively anaerobic bacteria the major pathway of carbohydrate uptake involves the phosphoenolpyruvate:sugar phospho-transferase system (PTS) (for a recent review, see ref. 54). The PTS is regulated by a variety of mechanisms that act at the levels of both enzyme synthesis and enzyme activity, and it is also of great importance in the control of peripheral metabolic pathways. Figure 4 illustrates the components involved in the operation of the PTS. Recent work has shown that protein phosphorylation is involved in the regulation of the PTS in the streptococci.

### A. DISCOVERY OF THE SYSTEM: INDUCER EXPULSION IN *STREPTOCOCCUS PYOGENES*

This work stemmed from investigation of the molecular basis of "inducer expulsion", a term that was first coined[55] to describe a phenomenon observed with *Streptococcus pyogenes*. In this organism, lactose and its non-metabolizable analogues such as thio-$\beta$-methylgalactoside (TMG) are taken up and phosphorylated via the PTS. Addition of TMG to cells therefore results in the accumulation of TMG-phosphate. Addition of glucose or mannose to cells pre-loaded with TMG-phosphate elicited an extremely rapid expulsion of free TMG.[55] Studies of the stoichiometry of the process showed that about five molecules of TMG were expelled per molecule of glucose taken up,[56] ruling out vectorial transphosphorylation as primarily responsible for the expulsion. Rather, the expulsion involved intracellular dephosphorylation of TMG-phosphate followed by efflux of the free TMG mediated by a transport protein.[56]

Use of inhibitors of glycolysis and alternative energy sources such as arginine showed that both glycolytic intermediates and provision of ATP were required for expulsion. Reizer et al.[56] then showed by $^{32}$P$_i$-labelling experiments that conditions that led to expulsion also resulted in the rapid phosphorylation of a small protein of apparent $M_r$ 10 000. This protein was subsequently identified[57] as HPr, the histidine-containing phospho-carrier protein of the PTS. HPr is transiently phosphorylated on a histidine residue during turnover of the PTS (Fig. 4), but the glucose-induced phosphorylation of HPr was on a serine residue.[57,58]

Reizer et al.[58] partially purified a protein kinase from Str. pyogenes that could transfer the γ-phosphoryl group of ATP to a serine residue of HPr. The kinase appeared to be membrane-bound in crude extracts but it was released in the presence of 1·5 M MgCl$_2$. It could phosphorylate the HPrs isolated from several streptococci and from Bacillus subtilis, but not that from E. coli. Phosphorylation of the serine residue of HPr was strongly inhibited by prior PEP-dependent phosphorylation of the active histidine residue. The kinase was activated by several intermediates of glycolysis and the pentose phosphate pathway, of which fructose 1,6-bisphosphate was the most potent.[58]

In this system, therefore, preferred sugars seem able to promote both expulsion of other sugars and serine-phosphorylation of HPr. However, it is not yet clear how (or indeed whether) the phosphorylation of HPr gives rise to sugar expulsion. One possibility is that P-Ser-HPr interacts with and activates a sugar-phosphate hydrolase that dephosphorylates the accumulated sugar-phosphate.[59] The intracellular concentrations of fructose 1,6-bisphosphate and other metabolites might play an important role in feedback regulation of sugar uptake by their effects on the protein kinase.[58,59]

## B. STUDIES WITH OTHER STREPTOCOCCI

Phosphorylation of HPr has been studied intensively in other streptococci. Enzymes responsible for the phosphorylation and dephosphorylation of a serine residue in HPr have been purified to homogeneity from Str. faecalis.[60,61] The kinase, a monomer of $M_r$ 65 000, was activated by fructose 1,6-bisphosphate and inhibited by P$_i$.[60] The site of phosphorylation has recently been identified as serine-46 of Str. faecalis HPr.[62] Deutscher et al.[61] purified a soluble HPr-phosphatase, which was stimulated by P$_i$, and also detected a membrane-bound activity which was not studied further.

Deutscher et al.[63] showed that P-Ser-HPr of Str. lactis was phosphorylated by Enzyme I of the PTS and PEP about 5000 times more slowly than was

HPr. However, the phosphorylation of P-Ser-HPr was increased in the presence of various Enzyme III components of the PTS. Thus the Enzyme III specific for lactose from *Staphylococcus aureus* increased the rate of phosphorylation 50-fold, while in the presence of the Enzyme III specific for gluconate from *Str. faecalis*, P-Ser-HPr was phosphorylated as rapidly as was HPr. The relevance of these experiments can be questioned because components of the PTS from different bacteria were used. However, it has been shown, using components of the lactose PTS from *Staph. aureus*, that the PEP-dependent phosphorylation of *o*-nitrophenyl-$\beta$-D-galactopyranoside is 150-fold slower in the presence of P-Ser-HPr than in the presence of HPr.[61] In combination, these data led to the proposal that phosphorylation of HPr on a serine residue is involved in an "inducer exclusion" mechanism in which the broad specificity of HPr is reduced so that it can discriminate amongst the Enzyme III components with the result that some PTS sugars are taken up in preference to others.[59,61,63]

HPr can also be phosphorylated on a serine residue by addition of ATP to extracts of the oral pathogen *Str. mutans*.[64] The rate of the reaction was enhanced by fructose 1,6-bisphosphate and reduced by prior incubation of the extract with PEP, which would be expected to cause phosphorylation of HPr on its active histidine residue. It was suggested that the role of HPr phosphorylation may be in feedback regulation of sugar uptake or in inducer exclusion.[64]

A general feature of the HPr kinases that have been studied to date is that they are activated by several metabolites, particularly fructose 1,6-bisphosphate. In addition, $P_i$ may be an important regulator of the system, either as an activator of HPr phosphatase[61] or as an inhibitor of HPr kinase.[60] It is known that starved streptococcal cells contain a low concentration of fructose 1,6-bisphosphate and a high concentration of $P_i$, while addition of glucose to such cells causes rapid accumulation of fructose 1,6-bisphosphate and a decline in $P_i$ concentration.[65–67]

These results are consistent with the view that the formation of P-Ser-HPr in Gram-positive organisms is mediated by the uptake of those carbohydrates whose metabolism can lead to a high intracellular concentration of fructose 1,6-bisphosphate,[61] and affects the functioning of the PTS. The effects of phosphorylation (e.g. inducer expulsion or inducer exclusion) may be different in different organisms. The PTS is clearly an extremely complex system, and more detailed experimental work will be required before this general hypothesis can be amplified and accepted. For example, the phosphorylation state of HPr must be assessed in intact cells metabolizing different sugars, and measurements of the rates of sugar uptake and of metabolite concentrations (particularly those of $P_i$ and fructose 1,6-bisphosphate) must be made in parallel. More studies are required of the effect of HPr phosphorylation on the function of the PTS for various sugars, using reconstituted systems. More details of the role and importance of the phosphorylation of HPr in Gram-positive bacteria are likely to emerge shortly.

Some recent observations suggest that another, unusual phosphorylation of a protein may play a role in the regulation of glycerol uptake and metabolism in streptococci. It has been shown that the PTS Enzyme I and HPr can catalyse PEP-dependent phosphorylation of the dihydroxyacetone kinase of *Str. faecalis*.[68,69] This enzyme can also phosphorylate glycerol, and its activity towards both substrates was increased some ten-fold by its phosphorylation.[69] The phosphorylated residue in the dihydroxyacetone kinase was shown to be a 3-phosphohistidine.[69] The phosphorylation of dihydroxyacetone kinase was considerably slower than that of the lactose-specific Enzyme III of the PTS, which competes for the common phosphoryl donor P-His-HPr. Therefore, it was suggested[69] that the phosphorylation of dihydroxyacetone kinase might occur only in the absence of a PTS sugar and might enhance the rate of glycerol or dihydroxyacetone metabolism in these conditions. The phosphorylation might therefore play a role similar to the allosteric regulation of glycerol kinase in Gram-negative bacteria by the glucose-specific Enzyme III of the PTS (e.g. ref. 54). If it can be shown that the phosphorylation of dihydroxyacetone kinase is significant in intact cells, this would be the first example of enzyme regulation by phosphorylation of a histidine residue.

## V. Regulation of Citrate Lyase Ligase in *Clostridium sphenoides*

Anaerobic utilization of citrate by microorganisms involves the reactions catalysed by citrate lyase and oxaloacetate decarboxylase as follows (reviewed in ref. 70)

$$\text{Citrate} \rightleftharpoons \text{oxaloacetate} + \text{acetate}$$

$$\text{Oxaloacetate} \rightarrow \text{pyruvate} + CO_2$$

Citrate lyase can exist in an active, acetylated form or an inactive, sulphydryl form. The acetyl groups in the active form are linked to a CoA-like prosthetic group and participate in the enzymic reaction:[70,71]

$$\text{Enzyme-}S\text{-acetyl} + \text{citrate} \rightarrow \text{Enzyme-}S\text{-citroyl} + \text{acetate}$$

$$\text{Enzyme-}S\text{-citroyl} \rightarrow \text{Enzyme-}S\text{-acetyl} + \text{oxaloacetate}$$

Acetylation of the inactive sulphydryl form of citrate lyase is catalysed *in vivo* by citrate lyase ligase.

The regulation of citrate lyase seems to depend on whether citrate synthase is also present. In organisms that lack citrate synthase (and hence require glutamate for growth) (e.g. *Clostridium sporosphaeroides*, *Leuconostoc citrovorum* and *Str. lactis*), neither citrate lyase nor citrate lyase ligase is regulated, presumably because no futile cycling between citrate and oxaloacetate is possible.[70] In contrast, other anaerobes (e.g. *Rhodocyclus gelatinosus* and *C. sphenoides*) can synthesize glutamate via citrate synthase. Citrate lyase is strictly controlled in these organisms, apparently to prevent competition between biosynthetic (citrate synthase) and degradative (citrate lyase) reactions.

In *R. gelatinosa* citrate lyase is activated by acetylation when citrate is present in the growth medium and inactivated by deacetylation upon exhaustion of the citrate.[71,72] After deacetylation the organism can synthesize glutamate from oxaloacetate and acetyl-CoA and can therefore grow on substrates other than citrate. In *C. sphenoides*, citrate lyase is activated by acetylation in the presence of citrate,[73] but deactivation in the absence of citrate seems to involve an unidentified modification prior to deacetylation.[74] Regulation of citrate lyase ligase almost certainly plays a significant role in the system; in both organisms this enzyme is rapidly inactivated in the absence of citrate.[73,75]

There is now evidence that citrate lyase ligase is regulated by reversible phosphorylation in *C. sphenoides*.[73] Exhaustion of citrate after growth on this compound resulted in a 90–95% decrease in the specific activity of citrate lyase ligase in crude extracts, as judged by assays at a low concentration of ATP. This partial inactivation could be reversed by addition of glutamate to the culture medium. The ligase was purified to apparent homogeneity from cells harvested either before or after the partial inactivation. Kinetic studies revealed that the partial inactivation was largely due to a ten-fold increase in the $K_m$ of the ligase for ATP, with little change in its $V_{max}$. This suggested that a reversible covalent modification might be involved. Isolation of the ligase from cultures labelled with $^{32}P_i$ showed that the active form contained $^{32}P$, but that the partially inactive form, obtained after exhaustion of citrate, did not. The nature of the $^{32}P$-containing group in the active form of the ligase has not yet been established. However, active ligase purified from cultures labelled with $[^{14}C]$inosine or $[^{14}C]$orotic acid contained no $^{14}C$-radioactivity. Treatment of the $^{32}P$-labelled ligase with acid phosphatase both reduced ligase activity and released $^{32}P$ from the protein. The existence of phosphorylated proteins in *C. sphenoides* has already been demonstrated,[19] although their identities have not been established. These results strongly suggest that reversible phosphorylation of citrate lyase ligase is involved in control of its activity. A similar system may occur in *R. gelatinosa*.[75] Clearly, many details remain to be resolved, including the phosphorylation stoichiometry *in vitro* and in intact cells and the properties of the relevant protein kinase and phosphatase, before the significance of the system can be fully appreciated.

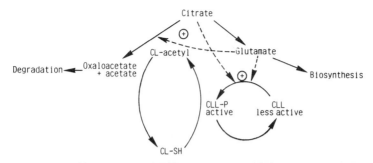

**Fig. 5.** Regulation of citrate metabolism in *Clostridium sphenoides* by acetylation and dephosphorylation. CL-acetyl and CL-SH represent the active and inactive forms of citrate lyase respectively. CLL represents citrate lyase ligase, which is activated by phosphorylation. The dotted lines indicate that the presence of citrate or glutamate in the growth medium give rise to phosphorylation of CLL, and that glutamate activates CL.

Figure 5 shows a tentative outline of the control of citrate lyase by acetylation and phosphorylation, and the likely role of this system in controlling the metabolic fate of citrate. The data currently available are compatible with the views that citrate lyase is activated in response to citrate or glutamate in the external medium and that this renders the degradation of citrate possible. In the absence of extracellular citrate and glutamate, citrate lyase is deacetylated and any intracellular citrate can be used for biosynthesis. It seems likely that the interconverting enzymes involved respond to the intracellular concentrations of some "signal" metabolites, but the nature of these is unknown. Citrate lyase is activated by glutamate *in vitro* but citrate lyase ligase is not.[75,76] Studies of the kinase and phosphatase responsible for the interconversions of citrate lyase ligase may help to reveal the nature of the "signals" and to clarify the role of the system.

## VI. Other Systems

Protein phosphorylation has been implicated in the regulation of several other processes in prokaryotes. One of the most fascinating examples concerns control of the synthesis of glutamine synthetase in *E. coli*. Transcription of the *glnALG* operon, which includes the structural gene *glnA* encoding glutamine synthetase, is controlled by nitrogen availability. This involves the DNA binding protein $NR_I$, which is encoded by *glnG*, and it can be controlled by a regulatory protein $NR_{II}$, the product of the gene *glnL* (e.g. ref. 77). Regulation of *glnA* expression by $NR_{II}$ also requires the products of the *glnB* and *glnD* genes.[78,79] The former encodes the regulatory protein $P_{II}$ involved in the control of glutamine synthetase by adenylation,

## TABLE 1

Some gene products involved in the regulation of the activity and expression of glutamine synthetase

| Gene | Product | Comments |
|------|---------|----------|
| glnA | Glutamine synthetase | |
| glnB | $P_{II}$ | Regulatory protein; the uridylated form stimulates deadenylation of glutamine synthetase and the deuridylated form stimulates adenylation. |
| glnD | Uridyltransferase | Catalyses both uridylation and deuridylation of $P_{II}$. |
| glnE | Adenyltransferase | Catalyses both adenylation and deadenylation of glutamine synthetase, controlled by $P_{II}$. |
| glnG | $NR_I$ | DNA binding protein; activates transcription from nitrogen-regulated promoters, see text. |
| glnL | $NR_{II}$ | Putative protein kinase/phosphatase, apparently regulated by $P_{II}$, see text. |

Ammonia starvation results in a high intracellular ratio of 2-oxoglutarate to glutamine and causes the conversion of $P_{II}$ to $P_{II}$-UMP.

and the latter codes for the uridyltransferase that mediates the interconversion between $P_{II}$ and $P_{II}$-UMP (e.g. ref. 80). The roles of some relevant gene products are summarized in Table 1. Genetic analysis showed that the effects of the glnB and glnD gene products on glnA expression are mediated by $NR_{II}$.[79] It has been proposed that, in the presence of $P_{II}$, $NR_{II}$ converts $NR_I$ to a form incapable of activating transcription at nitrogen-regulated promoters; when $P_{II}$ is converted to $P_{II}$-UMP, however, $NR_{II}$ converts $NR_I$ to a form capable of activating such transcription.

Ninfa and Magasanik[81] investigated the nature of the interaction between $NR_I$ and $NR_{II}$ using highly purified components *in vitro*. They found that $NR_{II}$ catalysed acid-stable incorporation of the $\gamma$-phosphoryl group of ATP into $NR_I$, to the extent of 0·45 groups per $NR_I$ dimer; concurrently, $NR_I$ became able to activate transcription from the glnAp2 promoter. When $[\alpha\text{-}^{32}P]$ATP was used, no incorporation of $^{32}P$ into $NR_I$ was detected. Thus, although the nature of the $^{32}P$-containing group in modified $NR_I$ has not been identified, the modification must be a phosphorylation or, less likely, a pyrophosphorylation.

Addition of the regulatory protein $P_{II}$ during the modification catalysed by $NR_{II}$ resulted in losses of both $^{32}P$ from $NR_I$ and the ability of $NR_I$ to activate transcription at the glnAP2 promoter. The effects of $P_{II}$-UMP have yet to be reported. $NP_{II}$ thus appears to possess both kinase and phosphatase activity against $NR_I$; it may be analogous to ICDH kinase/phosphatase (see Section III). Protein $P_{II}$ seems to affect two different parts of the glutamine synthetase system. The unmodified form of this protein, that present during nitrogen excess, stimulates the adenylation of glutamine synthetase and the apparent dephosphorylation of $NR_I$. The effects are that glutamine synthetase is inactivated and its synthesis is halted.

A major experimental advantage of the glutamine synthetase system is the wealth of genetic information that is available. One *glnL* mutant, namely *glnL2302*, encodes an $NR_{II}$ altered in such a way that transcription is activated at nitrogen-regulated promoters irrespective of nitrogen availability.[82,83] Ninfa and Magasanik[81] showed that the mutant protein $NR_{II}2302$ could phosphorylate $NR_I$ somewhat faster than could the wild-type protein, but that addition of $P_{II}$ did not promote dephosphorylation. This finding might suggest that the kinase and phosphatase active sites of $NR_{II}$ are distinct, and it might explain the biological effects of the mutation.

This is the first system in which protein phosphorylation has been implicated in the control of gene expression in a prokaryote. Similar phenomena may occur in other systems. Phosphorylation of DNA-binding proteins such as $NR_I$ could escape detection in intact cells because of the low abundance of these proteins. Measurements of the phosphorylation state of a low abundance protein in intact cells would probably require immunoprecipitation of the protein. Such measurements will be important in future work on the glutamine synthetase system. The significance of the phosphorylation of $NR_I$ must be assessed by measurements of the phosphorylation state of the protein in intact cells as a function of nitrogen availability. The prediction that $NR_I$ should be phosphorylated in the *glnL2302* mutant irrespective of nitrogen availability should be tested. Further biochemical work, in the context of the genetic information already available, should lead to rapid progress in our understanding of this system.

It is well established that the product of the *dnaK* gene of *E. coli* is phosphorylated in intact cells.[10,84] This protein, first described as a gene product necessary for bacteriophage $\lambda$-DNA replication, is a heat-shock protein that seems to play a role in termination of the heat-shock response.[85] Zylicz *et al.*[84] showed that the protein possessed a weak ATPase activity, could apparently self-phosphorylate on a threonine residue *in vitro*, and was phosphorylated in intact cells. The *dnaK* protein is homologous to the hsp70 heat-shock protein of eukaryotes (e.g. ref. 86); it is interesting to note that the hsp70 of *Dictyostelium discoideum* can also be phosphorylated on a threonine residue.[87] The *dnaK* protein is one of the most abundant proteins in *E. coli*,[88] yet in the $^{32}P_i$-labelling experiments of Cortay *et al.*[10] (see Section II) it gave only a faint spot on autoradiography. This suggests that only a very small proportion of the protein is phosphorylated in intact cells, at least in the conditions that have been tested so far. The biological significance of the phosphorylation is therefore open to question.

Studies of protein phosphorylation as a regulatory mechanism in pro-karyotes are now on the increase, and in several systems preliminary evidence implicating phosphorylation has been found. One example con-cerns the regulation of nitrogenase activity in *Chromatium vinosum*.[89] This is rapidly but reversibly inhibited by addition of $NH_4^+$ to the growth medium, mediated by inactivation of the Fe protein component of the nitrogenase. An experiment involving $^{32}P_i$-labelling of intact cells showed that the inactive Fe protein component contained $^{32}P$; the nature of the phosphate-containing group has yet to be identified. It has recently been reported that *E. coli* isocitrate lyase can be phosphorylated in crude extracts.[90] This observation is intriguing in view of the well-established role of phosphoryl-ation in the control of ICDH (Section III). The modification was acid-labile, and is presumably not of a serine, threonine or tyrosine residue. It is not yet known whether this modification affects the activity of the lyase.

## VII. Conclusions

In the eight years since its occurrence was first conclusively demonstrated, it has become clear that protein phosphorylation in prokaryotes is actually quite widespread. The implication is that protein phosphorylation is an important regulatory device in prokaryotes, perhaps even as important as it is in eukaryotes. However, a major difficulty in assessing the importance of phosphorylation in prokaryotes is that so few of the target proteins have been positively identified. It may be instructive to consider whether any general principles emerge from those systems in which the target protein has been identified, because such principles might facilitate identification of other substrates for phosphorylation.

Protein phosphorylation is involved in metabolic control (*E. coli* ICDH and *C. sphenoides* citrate lyase), control of gene expression (the glutamine synthetase system in *E. coli*) and transport processes (HPr and the PTS in some Gram-positive organisms). In metabolic control the target of phos-phorylation may be either an enzyme of primary metabolism (ICDH) or an interconverting enzyme (citrate lyase ligase). In spite of this diversity, two common themes emerge when the functions of the systems and the "signals" involved in their control are considered.

First, the function of each system seems to be to adjust the activities of key enzymes within the cell to make best use of the nutrients available in the growth medium. Secondly, the systems are controlled by the levels of one or (more frequently) several metabolites that seem to act as intracellular "signals". The effects of these metabolites may be mediated directly by the protein kinase and/or protein phosphatase involved, as in the examples of ICDH and HPr. In the glutamine synthetase system, by contrast, effects on the phosphorylation of $NR_I$ seem to be mediated indirectly, via the uridylation state of $P_{II}$.

In view of the suggestion (see Section II) that, at least in *E. coli*, several of the targets of phosphorylation are fairly abundant cellular proteins, it seems reasonable to speculate that these may include some regulatory enzymes of the central metabolic pathways other than ICDH. Changes in the degree of modification of an enzyme that is regulated by phosphorylation should be detected most readily if growth conditions can be manipulated in such a way as to alter the necessary flux through the enzyme in question. Experiments aimed at identifying targets for phosphorylation should be designed with this in mind. In retrospect, it can be seen that the experimental system in which the reversible inactivation of ICDH was discovered, namely growth on glucose and excretion of acetate followed by adaptation to and use of acetate, was almost the ideal one.

In all likelihood, some of the as yet unidentified targets of phosphorylation will prove to be proteins involved in other cases of transport or gene expression. Such targets are likely to be identified only if the possible involvement of a covalent modification is tested directly. If activation or inactivation of a protein's function is known to occur rapidly in intact cells, and in particular if the changes are reversible, it would seem appropriate to ask whether protein synthesis or degradation is involved. If not, protein phosphorylation would represent a real possibility, and its occurrence could be tested by, for example, immunoprecipitation of the protein from $^{32}P_i$-labelled cells.

Two phosphorylation systems in *E. coli* seem to involve bifunctional regulatory enzymes, namely ICDH kinase/phosphatase and the *glnL* gene product $NR_{II}$. Such enzymes are also found in eukaryotic systems, for example the phosphofructokinase-2/fructose-2,6-bisphosphatase of mammalian liver, and the regulatory protein of pyruvate, $P_i$ dikinase in maize (e.g. refs 91 and 92 respectively). The adenyltransferase and uridyl-transferase involved in the control of *E. coli* glutamine synthetase are other examples of bifunctional regulatory enzymes (e.g. ref. 80). It can be suggested that these systems have one property in common: the regulatory enzyme responds to a relatively large number of different metabolites, and effectively integrates a wide variety of metabolic information. In contrast, many of the monofunctional protein kinases in eukaryotic systems respond to a single compound, such as a cyclic nucleotide, $Ca^{2+}$ ions or diacylglycerol, and are involved in signal transduction. Perhaps the existence of a bifunctional regulatory enzyme represents a selective advantage in those systems that must respond to multiple intracellular signals. If this is the case, other examples of bifunctional regulatory enzymes may well be found in prokaryotes.

In summary, protein phosphorylation operates as a regulatory mechanism at several different levels to control various facets of prokaryotic cell functions, and more examples are likely to be uncovered in the near future. It is likely that studies of these systems will prove to be very important for our understanding of the functions and operation of covalent modification systems in general. There are many experimental advantages to be gained by investigating phosphorylation in well-studied bacteria. These include the use of genetics, the ease of gene cloning and the ability to control the level of expression of cloned genes. Experiments involving use of these techniques are already under way, and are likely to increase our understanding of the operation of phosphorylation systems substantially in the future.

## ACKNOWLEDGEMENTS

Work from this laboratory has been carried out in collaboration with W. H. Holms and many other colleagues to whom I express my thanks. Our research has been supported by the Science and Engineering Research Council. I hold a Senior Research Leave Fellowship awarded by the Medical Research Council.

## REFERENCES

1. Krebs, E. G. & Beavo, J. A. (1979). Phosphorylation–dephosphorylation of enzymes. *Annu. Rev. Biochem.* **48**, 923–959.
2. Hunter, T. & Cooper, J. A. (1985). Protein-tyrosone kinases. *Annu. Rev. Biochem.* **54**, 897–930.
3. Kuo, J. F. & Greengard, P. (1969). An adenosine 3′,5′-monophosphate-dependent protein kinase from *Escherichia coli. J. Biol. Chem.* **244**, 3417–3419.
4. Li, H.-C. & Brown, G. G. (1973). Orthophosphate and histone-dependent polyphosphate kinase from *E. coli. Biochem. Biophys. Res. Commun.* **53**, 875–881.
5. Rahmsdorf, H. J., Pai, S. H., Ponta, H., Herrlich, P., Roskoski, R., Schweiger, M. & Studier, F. W. (1974). Protein kinase induction in *Escherichia coli* by bacteriophage T7. *Proc. Natl. Acad. Sci. USA* **71**, 586–589.
6. Zillig, W., Fujiki, H., Blum, W., Janekovic, D., Schweiger, M., Rahmsdorf, H. J., Ponta, H. & Hirsch-Kauffmann, M. (1972). *In vivo* and *in vitro* phosphorylation of DNA-dependent RNA polymerase of *Escherichia coli* by bacteriophage-T7-induced protein kinase. *Proc. Natl. Acad. Sci. USA* **72**, 2506–2510.
7. Rubin, C. S. & Rosen, O. M. (1975). Protein phosphorylation. *Annu. Rev. Biochem.* **44**, 831–887.
8. Wang, J. Y. J. & Koshland, D. E., Jr (1978). Evidence for protein kinase activities in the prokaryote *Salmonella typhimurium. J. Biol. Chem.* **253**, 7605–7608.

9. Manai, M. & Cozzone, A. J. (1979). Analysis of the protein-kinase activity of *Escherichia coli* cells. *Biochem. Biophys. Res. Commun.* **91**, 819–826.
10. Cortay, J.-C., Rieul, C., Duclos, B. & Cozzone, A. J. (1986). Characterization of the phosphoproteins of *Escherichia coli* cells by electrophoretic analysis. *Eur. J. Biochem.* **159**, 227–237.
11. Garrison, J. C. & Wagner, J. D. (1982). Glucagon and the Ca$^{++}$-linked hormones angiotensin II, norepinephrine and vasopressin stimulate the phosphorylation of distinct substrates in intact hepatocytes. *J. Biol. Chem.* **257**, 13 135–13 143.
12. Desmarquets, G., Cortay, J. C. & Cozzone, A. J. (1984). Two-dimensional analysis of proteins phosphorylated in *E. coli* cells. *FEBS Lett.* **173**, 337–341.
13. Enami, M. & Ishihama, A. (1984). Protein phosphorylation in *Escherichia coli* and purification of a protein kinase. *J. Biol. Chem.* **259**, 526–533.
14. Cortay, J. C., Duclos, B. & Cozzone, A. J. (1986). Phosphorylation of an *Escherichia coli* protein at tyrosine. *J. Mol. Biol.* **187**, 305–308.
15. Borthwick, A. C., Holms, W. H. & Nimmo, H. G. (1984) Isolation of active and inactive forms of isocitrate dehydrogenase from *Escherichia coli* ML308. *Eur. J. Biochem.* **141**, 393–400.
16. Bennett, P. M. & Holms, W. H. (1975). Reversible inactivation of the isocitrate dehydrogenase of *Escherichia coli* ML308 during growth on acetate. *J. Gen. Microbiol.* **87**, 37–51.
17. Quentmeier, A. & Antranikian, G. (1986). Covalent modification of proteins in *Escherichia coli* growing anaerobically with nitrate as electron acceptor. *FEMS Microbiol. Lett.* **34**, 231–235.
18. Ferro-Luzzi Ames, G. & Nikaido, K. (1981). Phosphate-containing proteins of *Salmonella typhimurium* and *Escherichia coli*. *Eur. J. Biochem.* **115**, 525–531.
19. Antranikian, G., Herzberg, C. & Gottschalk, G. (1985). *In vivo* phosphorylation of proteins in *Clostridium sphenoides*. *FEMS Microbiol. Lett.* **27**, 135–138.
20. Averhoff, B., Antranikian, G. & Gottschalk, G. (1986). Phosphorylation and nucleotidylation of proteins in *Rhodocyclus gelatinosus*. *FEMS Microbiol. Lett.* **33**, 299–304.
21. Wang, J. Y. J. & Koshland, D. E., Jr (1981). The identification of distinct protein kinases and phosphatases in the prokaryote *Salmonella typhimurium*. *J. Biol. Chem.* **256**, 4640–4648.
22. Kornberg, H. L. (1966). Anaplerotic sequences and their role in metabolism. *Essays Biochem.* **2**, 1–31.
23. Ashworth, J. M. & Kornberg, H. L. (1963). Fine control of the glyoxylate cycle by allosteric inhibition of isocitrate lyase. *Biochim. Biophys. Acta* **73**, 519–522.
24. Lowry, O. H., Carter, J., Ward, J. B. & Glaser, L. (1971). The effect of carbon and nitrogen sources on the level of metabolic intermediates in *Escherichia coli*. *J. Biol. Chem.* **246**, 6511–6521.
25. Holms, W. H. & Bennett, P. M. (1971). Regulation of isocitrate dehydrogenase activity in *Escherichia coli* on adaptation to acetate. *J. Gen. Microbiol.* **65**, 57–68.
26. Garnak, M. & Reeves, H. C. (1979). Phosphorylation of isocitrate dehydrogenase of *Escherichia coli*. *Science* **203**, 1111–1112.
27. Garnak, M. & Reeves, E. C. (1979). Purification and properties of phosphorylated isocitrate dehydrogenase of *Escherichia coli*. *J. Biol. Chem.* **254**, 7915–7920.
28. LaPorte, D. C. & Koshland, D. E., Jr (1982). A protein with kinase and phosphatase activity involved in Krebs cycle regulation. *Nature (London)* **300**, 458–460.

29. Nimmo, G. A., Borthwick, A. C., Holms, W. H. & Nimmo, H. G. (1984). Partial purification and properties of isocitrate dehydrogenase kinase and isocitrate dehydrogenase phosphatase from *Escherichia coli* ML308. *Eur. J. Biochem.* **141**, 401–408.

30. LaPorte, D. C. & Chung, T. (1985). A single gene codes for the kinase and phosphatase which regulate isocitrate dehydrogenase. *J. Biol. Chem.* **260**, 15 291–15 297.

31. Brice, C. G. & Kornberg, H. L. (1968). Genetic control of isocitrate lyase activity in *Escherichia coli*. *J. Bacteriol.* **96**, 2185–2186.

32. Maloy, S. R. & Nunn, W. D. (1982). Genetic regulation of the glyoxylate shunt in *Escherichia coli* K-12. *J. Bacteriol.* **149**, 173–180.

33. LaPorte, D. C., Thorsness, P. & Koshland, D. E., Jr (1985). Compensatory phosphorylation of isocitrate dehydrogenase. *J. Biol. Chem.* **260**, 10 563–10 568.

34. El-Mansi, E. M. T., MacKintosh, C., Duncan, K., Holms, W. H. & Nimmo, H. G. (1987). Molecular cloning and over-expression of the glyoxylate bypass operon of *Escherichia coli* ML308. *Biochem. J.* **242**, 661–665.

35. LaPorte, D. C. & Koshland, D. E., Jr (1983) Phosphorylation of isocitrate dehydrogenase as a demonstration of enhanced sensitivity in covalent regulation. *Nature (London)* **305**, 286–290.

36. Borthwick, A. C., Holms, W. H. & Nimmo, H. G. (1984). Amino acid sequence round the site of phosphorylation in isocitrate dehydrogenase from *Escherichia coli* ML308. *FEBS Lett.* **174**, 112–115.

37. Malloy, P. J., Reeves, H. C. & Spiess, J. (1984). Amino acid sequence of the phosphorylation site of isocitrate dehydrogenase from *Escherichia coli*. *Curr. Microbiol.* **11**, 37–41.

38. Cortay, J.-C., Reeves, H. C. & Cozzone, A. J. (1986). Multiplicity of phosphorylation sites on *Escherichia coli* isocitrate dehydrogenase. *Curr. Microbiol.* **13**, 251–254.

39. Garland, D. & Nimmo, H. G. (1984). A comparison of the phosphorylated and unphosphorylated forms of isocitrate dehydrogenase from *Escherichia coli* ML308. *FEBS Lett.* **165**, 259–264.

40. Nimmo, H. G. (1984). Control of *Escherichia coli* isocitrate dehydrogenase: an example of protein phosphorylation in a prokaryote. *Trends Biochem. Sci.* **9**, 475–478.

41. Borthwick, A. C., Holms, W. H. & Nimmo, H. G. (1984). The phosphorylation of *Escherichia coli* isocitrate dehydrogenase in intact cells. *Biochem. J.* **222**, 797–804.

42. Wang, J. Y. J. & Koshland, D. E., Jr (1982). The reversible phosphorylation of isocitrate dehydrogenase of *Salmonella typhimurium*. *Arch. Biochem. Biophys.* **218**, 59–67.

43. Nimmo, G. A. & Nimmo, H. G. (1984). The regulatory properties of isocitrate dehydrogenase kinase and isocitrate dehydrogenase phosphatase from *Escherichia coli* ML308 and the roles of these activities in the control of isocitrate dehydrogenase. *Eur. J. Biochem.* **141**, 409–414.

44. Goldbeter, A. & Koshland, D. E., Jr (1981). An amplified sensitivity arising from covalent modification in biological systems. *Proc. Natl. Acad. Sci. USA* **78**, 6840–6844.

45. Reeves, H. C. & Malloy, P. J. (1983). Phosphorylation of isocitrate dehydrogenase in *Escherichia coli* mutants with a non-functional glyoxylate bypass. *FEBS Lett.* **158**, 239–242.

46. Walsh, K. & Koshland, D. E., Jr (1984) Determination of flux through the branch point of two metabolic cycles. *J. Biol. Chem.* **259**, 9646–9654.
47. LaPorte, D. C., Walsh, K. & Koshland, D. E., Jr (1984). The branch point effect. Ultrasensitivity and subsensitivity to metabolic control. *J. Biol. Chem.* **259**, 14068–14075.
48. Walsh, K. & Koshland, D. E., Jr (1985). Branch point control by the phosphorylation state of isocitrate dehydrogenase. *J. Biol. Chem.* **260**, 8430–8437.
49. Bautista, J., Satrustegui, J. & Machado, A. (1979). Evidence suggesting that the NADPH/NADP$^+$ ratio modulates the splitting of the isocitrate flux between the glyoxylic and tricarboxylic acid cycles in *Escherichia coli*. *FEBS Lett.* **105**, 333–336.
50. El-Mansi, E. M. T., Nimmo, H. G. & Holms, W. H. (1985). The role of isocitrate in control of the phosphorylation state of isocitrate dehydrogenase in *Escherichia coli* ML308. *FEBS Lett.* **183**, 251–255.
51. Nimmo, H. G. (1986). Kinetic mechanism of *Escherichia coli* isocitrate dehydrogenase and its inhibition by glyoxylate and oxaloacetate. *Biochem. J.* **234**, 317–323.
52. El-Mansi, E. M. T., Nimmo, H. G. & Holms, W. H. (1986). Pyruvate metabolism and the phosphorylation state of isocitrate dehydrogenase in *Escherichia coli*. *J. Gen. Microbiol.* **132**, 797–806.
53. Holms, W. H. (1986). The central metabolic pathways of *Escherichia coli*— relationship between flux and control at a branchpoint, efficiency of conversion to biomass and excretion of acetate. *Curr. Top. Cell. Regul.* **28**, 69–105.
54. Postma, P. W. & Lengeler, J. W. (1985). Phosphoenolpyruvate: carbohydrate phosphotransferase system of bacteria. *Microbiol. Rev.* **49**, 232–269.
55. Reizer, J. & Panos, C. (1980). Regulation of $\beta$-galactoside phosphate accumulation in *Streptococcus pyogenes* by an expulsion mechanism. *Proc. Natl. Acad. Sci. USA* **77**, 5497–5501.
56. Reizer, J., Novotny, M. J., Panos, C. & Saier, M. H., Jr (1983). Mechanism of inducer expulsion in *Streptococcus pyogenes*: a two-step process activated by ATP. *J. Bacteriol.* **156**, 354–361.
57. Deutscher, J. & Saier, M. H., Jr (1983). ATP-dependent protein kinase-catalyzed phosphorylation of a seryl residue in HPr, a phosphate carrier protein of the phosphotransferase system in *Streptococcus pyogenes*. *Proc. Natl. Acad. Sci. USA* **80**, 6790–6794.
58. Reizer, J., Novotny, M. J., Hengstenberg, W. & Saier, M. H., Jr (1984). Properties of ATP-dependent protein kinase from *Streptococcus pyogenes* that phosphorylates a seryl residue in HPr, a phosphocarrier protein of the phosphotransferase system. *J. Bacteriol.* **160**, 333–340.
59. Reizer, J., Deutscher, J., Sutrina, S., Thompson, J. & Saier, M. H., Jr (1985). Sugar accumulation in Gram-positive bacteria: exclusion and expulsion mechanism. *Trends Biochem. Sci.* **10**, 32–35.
60. Deutscher, J. & Engelmann, R. (1984). Purification and characterization of an ATP-dependent protein kinase from *Streptococcus faecalis*. *FEMS Microbiol. Lett.* **23**, 157–162.
61. Deutscher, J., Kessler, U. & Hengstenberg, W. (1985)-. Streptococcal phosphoenolpyruvate: sugar phosphotransferase system: purification and characterization of a phosphoprotein phosphatase which hydrolyzes the phosphoryl bond in seryl-phosphorylated histidine-containing protein. *J. Bacteriol.* **163**, 1203–1209.

62. Deutscher, J., Pevec, B., Beyreuther, K., Kiltz, H.-H. & Hengstenberg, W. (1986). Streptococcal phosphoenolpyruvate–sugar phosphotransferase system. Amino acid sequence and site of ATP-dependent phosphorylation of HPr. *Biochemistry* **25**, 6543–6551.

63. Deutscher, J., Kessler, U., Alpert, C. A. & Hengstenberg, W. (1984). Bacterial phosphoenolpyruvate-dependent phosphotransferase system: P-Ser-HPr and its possible regulatory function. *Biochemistry* **23**, 4455–4446.

64. Mimura, C. S., Poy, F. & Jacobson, G. R. (1986). ATP-dependent protein kinase activities in the oral pathogen *Streptococcus mutans*. *J. Cell. Biochem.* **33**, 161–171.

65. Thompson, J. & Torchia, D. A. (1984). Use of $^{31}$P nuclear magnetic resonance spectroscopy and $^{14}$C fluorography in studies of glycolysis and regulation of pyruvate kinase in *Streptococcus lactis*. *J. Bacteriol.* **158**, 791–800.

66. Yamada, T. & Carlsson, J. (1975). Regulation of lactate dehydrogenase and change of fermentation products in streptococci. *J. Bacteriol.* **124**, 55–61.

67. Mason, P. W., Carbone, D. P., Cushman, R. A. & Waggoner, A. S. (1981). The importance of inorganic phosphate in regulation of energy metabolism of *Streptococcus lactis*. *J. Biol. Chem.* **256**, 1861–1866.

68. Deutscher, J. (1985). Phosphoenolpyruvate-dependent phosphorylation of a 55-kDa protein of *Streptococcus faecalis* catalyzed by the phosphotransferase system. *FEMS Microbiol. Lett.* **29**, 237–243.

69. Deutscher, J. & Sauerwald, H. (1986). Stimulation of dihydroxyacetone and glycerol kinase activity in *Streptococcus faecalis* by phosphoenolpyruvate-dependent phosphorylation catalyzed by enzyme I and HPr of the phosphotransferase system. *J. Bacteriol.* **166**, 829–836.

70. Gottschalk, G., Giffhorn, F. & Antranikian, G. (1982). The regulation of citrate lyase by acetylation/deacetylation. *Biochem. Soc. Trans.* **10**, 324–326.

71. Giffhorn, F. & Gottschalk, G. (1975). Effect of growth conditions on the activation and inactivation of citrate lyase of *Rhodopseudomonas gelatinosa*. *J. Bacteriol.* **124**, 1046–1051.

72. Giffhorn, F. & Gottschalk, G. (1975). Inactivation of citrate lyase from *Rhodopseudomonas gelatinosa* by a specific deacetylase and inhibition of this inactivation by L-(+)-glutamate. *J. Bacteriol.* **124**, 1052–1061.

73. Antranikian, G., Herzberg, C. & Gottschalk, G. (1985). Covalent modification of citrate lyase ligase from *Clostridium sphenoides* by phosphorylation/dephosphorylation. *Eur. J. Biochem.* **153**, 413–420.

74. Herzberg, C. & Antranikian, G. (1986). Evidence for a unique pattern of citrate lyase inactivation in *Clostridium sphenoides*. *Biochim. Biophys. Acta* **871**, 107–120.

75. Antranikian, G., Giffhorn, F. & Gottschalk, G. (1978) Activation and inactivation of citrate lyase ligase from *Rhodopseudomonas gelatinosa*. *FEBS Lett.* **88**, 67–70.

76. Antranikian, G., Klinner, C., Kummel, A., Schwanitz, D., Zimmerman, T., Mayer, F. & Gottschalk, G. (1982). Purification of L-glutamate-dependent citrate lyase from *Clostridium sphenoides* and electron microscopic analysis of citrate lyase isolated from *Rhodopseudomonas gelatinosa*, *Streptococcus diacetilactis* and *C. sphenoides*. *Eur. J. Biochem.* **126**, 35–42.

77. Magasanik, B. & Bueno, R. (1985). The role of uridylyltransferase and $P_{II}$ in the regulation of the synthesis of glutamine synthetase in *Escherichia coli*. *Curr. Top. Cell. Regul.* **27**, 215–220.

78. Bloom, F. R., Levin, M. S., Foor, F. & Tyler, B. (1978). Regulation of glutamine synthetase formation in *Escherichia coli*: characterization of mutants lacking the uridyltransferase. *J. Bacteriol.* **134**, 569–577.
79. Bueno, R., Pahel, G. & Magasanik, B. (1985). Role of *glnB* and *glnD* gene products in regulation of the *glnALG* operon of *Escherichia coli*. *J. Bacteriol.* **164**, 816–822.
80. Rhee, S. G., Park, S. C. & Koo, J. H. (1985). The role of adenylyltransferase and uridylyltransferase in the regulation of glutamine synthetase in *Escherichia coli*. *Curr. Top. Cell. Regul.* **27**, 221–232.
81. Ninfa, A. J. & Magasanik, B. (1986). Covalent modification of the *glnG* product, $NR_I$, by the *glnL* product, $NR_{II}$, regulates the transcription of the *glnALG* operon in *Escherichia coli*. *Proc. Natl. Acad. Sci. USA* **83**, 5909–5913.
82. Chen, Y.-M., Backman, K. & Magasanik, B. (1982). Characterization of a gene, *glnL*, the product of which is involved in the regulation of nitrogen utilization in *Escherichia coli*. *J. Bacteriol.* **150**, 214–220.
83. Pahel, G., Zelenetz, A. P. & Tyler, B. M. (1978). *gltB* Gene and regulation of nitrogen metabolism by glutamine synthetase in *Escherichia coli*. *J. Bacteriol.* **133**, 139–148.
84. Zylicz, M., LeBowitz, J. H., McMacken, R. & Georgopoulos, C. (1983). The *dnaK* protein of *Escherichia coli* possesses an ATPase and autophosphorylating activity and is essential in an *in vitro* DNA replication system. *Proc. Natl. Acad. Sci. USA* **80**, 6431–6435.
85. Tilly, K., McKittrick, N., Zylicz, M. & Georgopoulos, C. (1983). The *dnaK* protein modulates the heat-shock response of *Escherichia coli*. *Cell* **34**, 641–646.
86. Burdon, R. H. (1986). Heat shock and the heat shock proteins. *Biochem. J.* **240**, 313–324.
87. Loomis, W. F., Wheeler, S. and Schmidt, J. A. (1982). Phosphorylation of the major heat shock protein of *Dictyostelium discoideum*. *Mol. Cell. Biol.* **2**, 484–489.
88. Herendeen, S. L., Van Bogelen, R. A. & Neidhardt, F. C. (1979). Levels of major proteins of *Escherichia coli* during growth at different temperatures. *J. Bacteriol.* **139**, 185–194.
89. Gotto, J. W. & Yoch, D. C. (1985). Regulation of nitrogenase activity by covalent modification in *Chromatium vinosum*. *Arch. Microbiol.* **141**, 40–43.
90. Robertson, E. F., Hoyt, J. C. & Reeves, H. C. (1987). *In vitro* phosphorylation of *Escherichia coli* isocitrate lyase. *Curr. Microbiol.* **15**, 103–105.
91. Pilkis, S. J., Regen, D. M., Stewart, B. H., Chrisman, T., Pilkis, J., Kountz, P., Pate, T., McGrane, M., El-Maghrabi, M. R. & Claus, T. H. (1984). Rat liver 6-phosphofructo-2-kinase/fructose-2,6-biphosphatase: a unique bifunctional enzyme regulated by cyclic AMP-dependent phosphorylation. In *Molecular Aspects of Cellular Regulation* (Cohen, P., ed.), Vol. 3, pp. 95–122. Elsevier, Amsterdam.
92. Burnell, J. N. & Hatch, M. D. (1985). Light-dark modulation of leaf pyruvate, $P_i$ dikinase. *Trends Biochem. Sci.* **10**, 288–291.

# Enzyme-activated/Mechanism-based Inhibitors

MICHAEL G. PALFREYMAN, PHILIPPE BEY and
ALBERT SJOERDSMA

*Merrell Dow Research Institute, 2110 East Galbraith Road,
Cincinnati, Ohio 45215, USA*

ESSAYS IN BIOCHEMISTRY Vol. 23
ISBN 0 12 158123-3

*Abbreviations*

| | | | |
|---|---|---|---|
| AADC | L-aromatic amino acid decarboxylase | GAD | glutamic acid decarboxylase |
| | | GAG | $\gamma$-acetylenic-GABA |
| D$\beta$H | dopamine-$\beta$-hydroxylase | 5-HTP | 5-hydroxytryptophan |
| DOPA | 3,4-dihydroxyphenylalanine | HDC | histidine decarboxylase |
| Dopamine | 3,4-dihydroxyphenethyl-amine | MAO | monoamine oxidase |
| | | MFMD | $\alpha$-monofluoromethyl-DOPA |
| DFMO | $\alpha$-difluoromethylornithine | MFMH | $\alpha$-monofluoromethyl-histidine |
| EOS | ethanolamine-$O$-sulphate | MFM-Tyr | $\alpha$-monofluoromethyl-tyrosine |
| FdUMP | 5-fluoro-2'-denoxyuridylic acid | MFM-Trp | $\alpha$-monofluoromethyl-tryptophan |
| FMMT | $\beta$-fluoromethylene-*meta*-tyrosine | Mo | molybdenum |
| | | ODC | ornithine decarboxylase |
| 5-FU | 5-fluorouracil | PAO | polyamine oxidase |
| GABA | 4-aminobutyric acid ($\gamma$-amino-butyric acid) | PLP | pyridoxal phosphate |
| | | XO | xanthine oxidase |
| GABA-T | 4-aminobutyric acid: 2-oxoglutarate aminotransferase | | |

# I. Introduction

To explore biological systems, biochemists and molecular pharmacologists require specific tools such as enzyme inhibitors. A major advance over the last decade and a half has been the discovery and development of enzyme-activated, mechanism-based inhibitors which have considerably improved the specificity and, hence, value of such research tools.

The approaches that have been taken to design and develop these new inhibitors, their implications for biological research and, in certain cases, their therapeutic use will be the subject of this chapter.

## II. What Constitutes an Enzyme-activated or Mechanism-based Inhibitor?

### A. NOMENCLATURE

Before starting to define the concepts, the problem of nomenclature should be addressed. Many authors[1-4] referred to these inhibitors as "suicide substrates" or "suicide inactivators". We have also been guilty of this error.[5] Suicide is a self-inflicted *voluntary* and *intentional* act leading to death.[6] As we shall see, enzymes do not commit suicide. They are in fact susceptible to terrorist attacks with "letter bombs". These are harmless until opened by the addressee. Rando used the more accurate, but rather non-descriptive term "$k_{cat}$ inhibitor"[7] and Merrell's scientists coined the term "enzyme-activated inhibitor".[8] Today the term "mechanism-based inhibitor" appears to be most widely used.[9-11] What, then, are the criteria for defining a compound as a mechanism-based inhibitor?

### B. DEFINITION OF A MECHANISM-BASED INHIBITOR

Any inhibitor that requires enzyme processing before it will inhibit that same enzyme is a mechanism-based inhibitor. Usually this type of inhibitor is a substrate or product analogue which upon turnover by the target enzyme leads to the formation of an intermediate that eventually inactivates the enzyme. In the vast majority of cases, the intermediate generated in the active site is an alkylating species which will form a subsequent covalent linkage with a residue from the active site. In some cases, this intermediate can be a tight binding ligand that will dissociate so slowly that, for all practical purposes, irreversible inhibition would result, although no covalent linkage is formed. Clearly the so-called "transition state" analogues are also "mechanism-based" inhibitors as they mimic one of the transition states along the reaction pathway catalysed by the enzyme. In principle, however, no activation by the target enzyme is required with this type of inhibitor. As implied, we will not deal with transition state inhibitors, but the reviews on this topic by Wolfenden[12] and Brodbeck[13] are recommended. To be perfectly logical, the "ideal" name for the inhibitors discussed in this review should be enzyme-activated/mechanism-based inhibitors.

### C. CRITERIA FOR ESTABLISHING THAT A COMPOUND IS A MECHANISM-BASED INHIBITOR

The key characteristic of enzyme-activated/mechanism-based inhibitors is that they are turned over by the target enzymes. In the first approximation, the kinetics of inactivation can be accounted for by the minimum scheme represented in Eqn 1. In most cases, however, the inhibitor can be

$$E + S' \underset{k_{-1}}{\overset{k_1}{\rightleftharpoons}} E{\cdot}S' \overset{k_{cat}}{\rightleftharpoons} E{\cdot}I \overset{k_{int.}}{\rightleftharpoons} E - I \tag{1}$$

turned over many times before inactivation occurs. The kinetic scheme of inactivation is then best described by Eqn 2:

$$
\begin{array}{c}
E + I \\
{\scriptstyle k_5} \Updownarrow \\
E + S' \underset{k_{-1}}{\overset{k_1}{\rightleftharpoons}} E{\cdot}S' \overset{k_{cat}}{\rightleftharpoons} E{\cdot}I \overset{k_{inh}}{\rightleftharpoons} E - I \\
{\scriptstyle k_4} \Updownarrow \\
E{\cdot}P' \rightleftharpoons E + P'
\end{array}
\tag{2}
$$

No single criterion is available to establish that the inhibitor is mechanism-based or to distinguish between mechanism-based inhibitors and affinity labelling agents whose minimum kinetic scheme for inactivation of an enzyme is represented in Eqn 3. Affinity labelling agents are irreversible inhibitors which *do not require* activation by the target enzyme.

$$E + I \underset{k_{-1}}{\overset{k_1}{\rightleftharpoons}} E{\cdot}I \overset{k_{inh}}{\longrightarrow} E - I \tag{3}$$

The demonstration that the inhibition is mechanism-based relies on a combination of kinetic and chemical criteria. The kinetic criteria are essentially identical to those used to characterize affinity labelling agents. Only chemical criteria, namely the determination of the structure of the $E - I$ adduct and the demonstration that the inhibitor is chemically transformed by the enzyme before inactivation occurs, can provide definitive evidence for mechanism-based inhibition.

### (1) Kinetic criteria

#### (a) Saturation kinetics

Assuming that the reactions depicted by Eqns 1 and 3 are studied under steady-state conditions, i.e. when $[E] \ll [I]$ or $[S']$, and that $k_{inh}$ or $k_{cat}$ are smaller than $k_1$ and $k_{-1}$, it can readily be established that in both cases the rate of enzyme inhibition will follow pseudo first-order kinetics. As the first step in both reactions is fully reversible, the law of mass action predicts that as the concentration of $I$ or $S'$ increases, a greater quantity of $E$ is present in the $E{\cdot}I$ or $E{\cdot}S'$ form and, hence, the rate of inactivation increases. Therefore, a semilog plot of the remaining enzyme activity against time as a

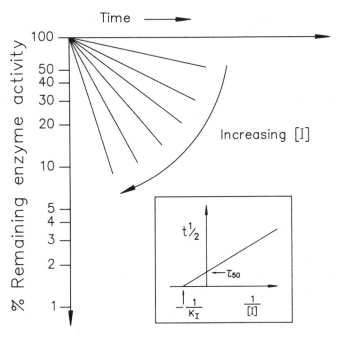

**Fig. 1.** Semilog plot of remaining enzyme activity against time as a function of various inhibitor concentrations.

function of various inhibitor concentrations yields a family of curves as illustrated in Fig. 1. From these curves, the time required to inactivate half of the enzyme can be determined for each concentration of inhibitor. By replotting the time to half inactivation as a function of the reciprocal of the inhibitor concentration as described by Kitz and Wilson,[14] a straight line is obtained. The intercept with the y axis corresponds to the $\tau_{50}$ or the minimum time to half inactivation under a saturating concentration of inhibitor, and the x axis intercept is equal to $-1/K_I$, $K_I$ being the apparent dissociation constant of the inhibitor. The slowest rate in the inactivation process, which in the minimum schemes represented by Eqns 1 and 3 could be either $k_{cat}$ or $k_{inh}$, can be calculated from Eqn 4:[15]

$$k_{inh} = 0 \cdot 69/\tau_{50} \tag{4}$$

There are many examples of mechanism-based inhibition where kinetics of inactivation are not linear.[16] The causes for the departure from linearity could be multiple. If crude enzyme preparations are used, the inhibitor could be metabolized by another enzyme in the preparation thus reducing the effective concentration of inhibitor. Non-linear kinetics are also observed with apparently pure enzyme preparations. In this case, the explanation could be the existence of isoenzymes with different sensitivities to inhibitors. High ratio between product formed and enzyme inactivation (the partition ratio) as illustrated by Eqn 2 or inactivation mechanisms involving multiple steps having rate constants within the same range invariably lead to complex kinetics which require more elaborate computer-aided solutions.[17] The Kitz and Wilson analysis is no longer applicable when kinetics are non-linear. Saturation can therefore no longer be determined directly from the inactivation curves. However, when the target enzyme has a high specific activity and the inhibitor a high affinity for the enzyme, $K_I$ can often be determined by using Lineweaver-Burk or Dixon plots.

On occasion, the Kitz and Wilson transformation can give a plot with the intercepts so close to the origin that neither $K_I$ nor $\tau_{50}$ can be determined. This situation often arises when the rate of inactivation is fast. Lowering the temperature of the assay may slow the inhibition step sufficiently to make it rate-determining.

Notwithstanding the problems, kinetic analysis *cannot* distinguish an affinity labelling agent from a mechanism-based inhibitor.

### (b) Protection against inactivation

The first step in the reaction, the binding to the enzyme active site, is common for the inhibitor and the natural substrate. It follows that normal substrates, or any ligand having an affinity for the active site, will compete with the inhibitors and therefore slow down the inactivation. This effect is illustrated in Fig. 2. This is a necessary, but not specific, criterion of mechanism-based inhibition.

### (c) Lag time, nucleophile addition and partition ratio

As depicted in Eqn 2, the activated intermediate $I$ has the possibility to diffuse from the active site. Subsequently $I$ could alkylate a different part of the protein to cause the inactivation of the enzyme or form a covalent bond with any other protein in the biological matrix. Such an inhibition is therefore more of the affinity-labelling than of the mechanism-based type. If such a mechanism is operative, then a lag time is likely to be observed as the concentration of $I$ has to build up before inactivation can be measured. This mechanism of inactivation can be ruled out or confirmed by two simple kinetic experiments. The rate of enzyme inactivation should be reduced dramatically when the enzyme assay is carried out in the presence of a nucleophile such as mercaptoethanol which would scavenge the activated species $I$. Secondly, the rate of inactivation of an additional sample of the enzyme added to an inactivated preparation should remain the same in the case of a true mechanism-based inhibitor but would be faster if

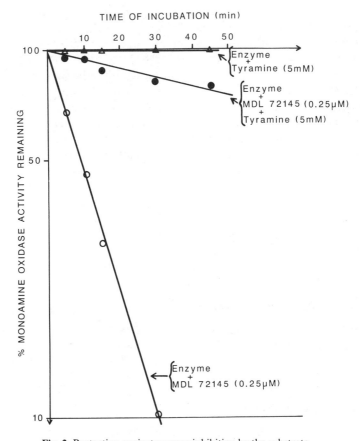

**Fig. 2.** Protection against enzyme inhibition by the substrate.

an activated species released in the medium is responsible for the inhibition.

As far as practical applications are concerned, the release of toxic "metabolites" is a real and troublesome possibility for mechanism-based inhibitors. As suggested in Eqn 2, mechanism-based inhibitors can be turned over many times before inactivation ensues. Thus, the ideal mechanism-based inhibitor should kill the target enzyme every time it is turned over. Such an inhibitor is said to have a partition ratio $(k_4 + k_5)/k_{inh}$ equal to one.

*(d) Stoichiometry*

A mechanism-based inhibitor should inactivate stoichiometrically the target enzyme. Normally, there are as many bound molecules of inhibitor as active sites present on the inhibited enzyme. This relationship has been exploited to titrate enzyme molecules (see p. 68). Occasionally, interaction of the inhibitors with one or two subunits of the enzyme may alter the conformation of the other subunits to block activity.[18] More importantly, a stoichiometry greater than the number of active sites suggests non-specific reaction.[19]

### (2) Chemical criteria

The *sine qua non* of mechanism-based inhibitors is their *absolute* requirement for catalytic processing by the target enzyme *prior* to inactivation. This single requirement gives this group of inactivators their unique advantages over other enzyme inhibitors. Verification of this crucial point is often difficult. However, enzyme turnover can be manipulated, e.g. by changes in temperature, cofactors, phosphorylation state of the enzyme or allosteric modulation of the activity, and all of these changes should have predictable repercussions on the *rate* of inactivation of the enzyme. In certain cases (e.g. with enzymes which abstract hydrogen), demonstration of a deuterium or tritium isotope effect may be a sufficiently convincing indication of catalytic turnover.[20,21] The stereospecificity (if it exists) of the inactivation process can also indicate an enzymatic turnover although its absence is not proof to the contrary. Release of a product formed during the activation process, such as carbon dioxide from a mechanism-based carboxylic acid inhibitor of a decarboxylase, is convincing proof of enzymatic processing.[22,23] Ideally, identification of intermediate products of the enzyme-mediated turnover of the inhibitor may be useful, but the best proof of a mechanism-based mode of inactivation is often the structural identification of the enzyme–inhibitor adducts. When the product of enzyme turnover is chemically stable, it is sometimes possible to inactivate directly the enzyme with this compound. This was achieved by Bloch and his co-workers[20,24] for the inhibition of $\beta$-hydroxydecanoyl thioester dehydrase. Unfortunately, the intermediates are rarely stable and frequently lack affinity for the enzyme; hence, complete resolution of the different steps involved in the mechanism-based inactivation are difficult to unravel in the majority of cases. Examples where enzyme–inhibitor adducts have been identified will be discussed later.

## D. POTENTIAL ADVANTAGES OF MECHANISM-BASED INHIBITORS

### (1) Specificity

Because mechanism-based inhibitors are chemically unreactive, they should inhibit *only* those enzymes that activate the molecule. However, there are many examples where enzymes generate reactive intermediates but do not become "inactivated", probably because the active site does not contain a suitably positioned reactive group. In spite of the impressive selectivity of many of these mechanism-based inhibitors, it should be stressed that specificity is constrained by nature's own limits of error. A pseudo-substrate does not have to be a good substrate to inactivate an enzyme efficiently if it is able at each turnover to kill one enzyme molecule. In fact, many mechanism-based inhibitors are extremely poor substrates of their target enzymes. This fact explains why, occasionally, enzymes are inhibited by pseudo-substrates which by the normal criteria of substrate specificities should not expect to be harmful to the enzyme. It is important that these properties can also be exploited in inhibitor design.

Stereospecificity of enzyme catalysis also holds a few surprises. Most enzyme-catalysed reactions are stereospecific. Again, one would expect the "right" isomer of the substrate analogue to be the inactivator and the "wrong" isomer to be inactive. As we shall see later, this is not always true. However, as a corollary to their inherent specificity, mechanism-based inhibitors are normally less "toxic" to the biological test system under investigation whether it be cells in culture, an animal, or a human patient.

## (2) *Duration of action*

Because the majority of the mechanism-based inhibitors forms *irreversible* adducts with the enzyme, activity will only return when new enzyme is resynthesized. Thus, the duration of action of such inhibitors does not depend on the biological half-life or pharmacokinetic attributes of the molecule, but rather on the biological half-life of the target enzyme. Naturally, in living cells or organisms, delivering the mechanism-based inhibitor to the target enzyme may be a major hurdle. This aspect will be more fully explored in Section XI.

## III. Historical Perspectives

It is now more than 15 years since Bloch and his co-workers[20] reported the fatal "error" that $\beta$-hydroxydecanoyl thioester dehydrase made when it tried to turnover 3-decynoyl-*N*-acetylcysteamine, an acetylenic analogue of its natural substrate. The original observation of Bloch and the further elucidation of the nature of the mechanism that led to the irreversible inhibition of the enzyme[20,24] represents a cornerstone in the development of the concept of mechanism-based inhibition and, as such, deserves to be discussed in detail.

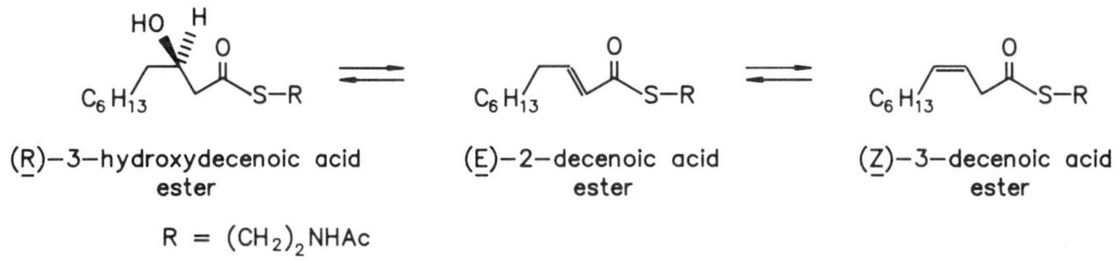

(R)-3-hydroxydecenoic acid          (E)-2-decenoic acid          (Z)-3-decenoic acid
ester                                ester                        ester

R = (CH$_2$)$_2$NHAc

SCHEME 1

The enzyme, $\beta$-hydroxydecanoyl thioester dehydrase of *Escherichia coli*, catalyses a series of reactions leading to an equilibrium between three products: D(-)-$\beta$-hydroxydecanoyl-CoA, (Z)-3-decanoyl-CoA, and (E)-2-decanoyl-CoA. It is known to use the corresponding N-acetylcysteamine derivatives as substrates as shown in Scheme 1.

This enzyme is pivotal in the anaerobic biosynthesis of unsaturated fatty acids and was shown by Bloch and co-workers[20,24-26] to be irreversibly inactivated by 3-decynoyl-N-acetylcysteamine as illustrated in Scheme 2.

The evidence indicative of mechanism-based inactivation can be summarized as follows:

(a) Inactivation of the enzyme is dependent on the time of incubation and concentration of the inhibitor.

(b) Loss of enzyme activity follows pseudo first-order kinetics and is irreversible.

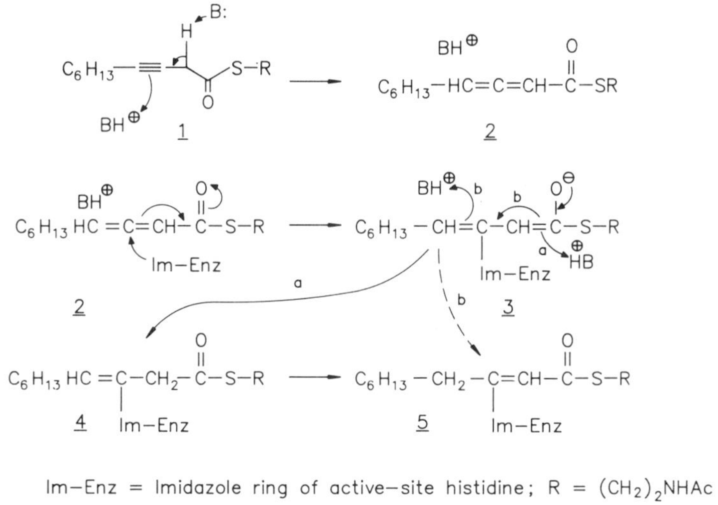

Im—Enz = Imidazole ring of active—site histidine; R = (CH$_2$)$_2$NHAc

SCHEME 2

(c) The rate of inactivation is slower with the 2,2-dideuterated analogue with an isotope effect $K_H/K_D = 2·6$. This isotope effect which is similar in magnitude to that seen in the natural substrate is consistent with abstraction of the $\alpha$ proton being the rate-limiting step in the enzymatic reaction.

(d) The rate of inactivation by the inhibitor is slowed down by the presence of the substrate.

(e) The nucleophile in the enzyme active site that is alkylated has been identified as histidine (3).

(f) The stoichiometry of the inactivation is 1 mol of (+)-2,3-decadienoyl-*N*-acetylcysteamine (2) for 1 mol of enzyme.

Unequivocal proof as to the identity of the enzyme–inhibitor adduct only became available in 1986. Elegant studies by Schwab and colleagues[27] using [$^{13}$C]NMR measurements on the substrate 3-[2-$^{13}$C]-decynoyl-*N*-acetyl-cysteamine with purified enzyme established clearly that the catalytic histidine residue adds to the 2,3-decadienoyl-*N*-acetylcysteamine to form initially the product (3-imidazolyl-3-decanoyl)-*N*-acetylcysteamine (4). This adduct then slowly isomerizes to the (3-imidazolyl-2-decanoyl)-*N*-acetyl-cysteamine (5) as shown in Scheme 3.

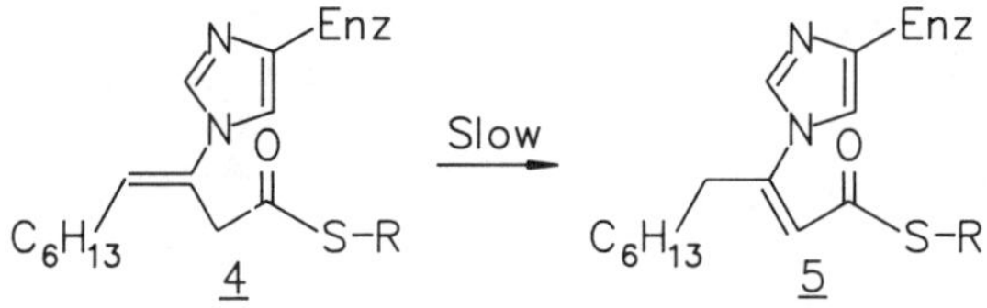

SCHEME 3

This mode of inactivation is completely consistent with the normal mode of action of the enzyme. Thus, the conversion of the acetylenic pseudo-substrate to the allene corresponds to a partial dehydrase reaction. The electrophilic 2,3-allene thioester intermediate then rapidly alkylates the histidine residue in the active site. The stereochemistry of the reaction sequence is also informative. Since the inhibitory allene has the $(S)$ configuration and the active site histidine removes (or supplies) the pro-2$(S)$ proton, the configuration of the double bond in the enzyme–inhibitor adduct should be $(E)$. This prediction illustrated in Scheme 4 was confirmed experimentally.

Since Bloch's key observation and the realization of its biological significance, a large number of enzyme-activated/mechanism-based inhibitors has been rationally designed in the last decade. Moreover, many "old" enzyme

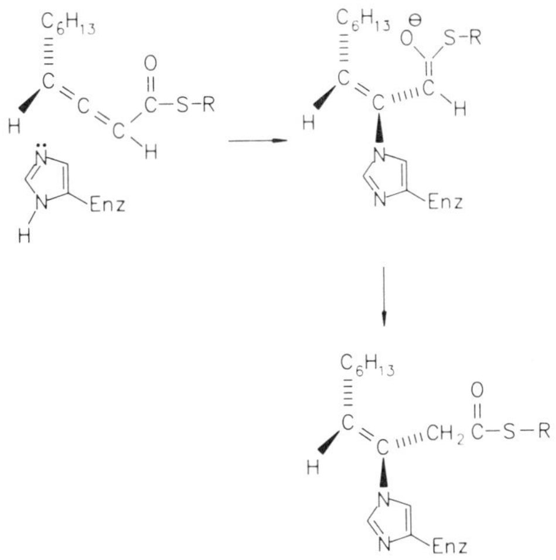

SCHEME 4

inhibitors, either of natural origin or the fruit of synthetic chemists' efforts, have been shown to owe their enzyme inhibitory property to a mechanism-based mode of action.

## IV. Design of Mechanism-based Inhibitors

As discussed in the preceding section, the formation of a carbanionic intermediate during turnover of the substrate constitutes the Achilles' heel of β-hydroxydecanoyl thioester dehydrase for its inactivation by the acetylenic mechanism-based inhibitor. Knowledge of the mechanism of action of the target enzyme is obviously of paramount importance to the design of mechanism-based inhibitors. Usually the nature of the intermediate generated during catalysis determines which approach to use to design a mechanism-based inhibitor. This topic has been reviewed many times with a focus either on the type of latent functional group,[28–30] the class of enzyme to be inhibited,[8,10,31,32] or the chemical reactions that are involved in the mechanism-based inactivation.[9] We have not attempted a comprehensive classification, but rather have selected examples to illustrate aspects of inhibitor design or uses which we feel important. This selection is inevitably influenced by our own experience. The reader interested in a more comprehensive description of mechanism-based inhibitors should refer to the excellent reviews by Seiler et al.,[8] Abeles and Maycock,[2] Walsh,[1,28] Silverman and Hoffman,[9] Rando,[10] and Kalman.[11]

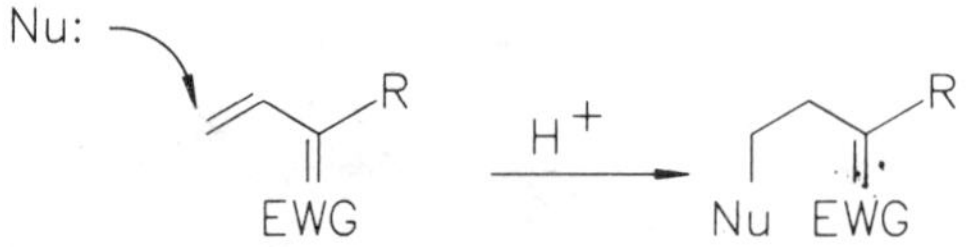

EWG = Electron withdrawing group which can be part of enzyme or cofactor.

Nu: = Nucleophilic residue in enzyme active site.

SCHEME 5

Pyridoxal-phosphate (PLP) and flavin-dependent enzymes are involved in the metabolism of a variety of physiologically important amino acids and biogenic amines. They have attracted a lot of attention as targets for inactivation by mechanism-based inhibitors. These enzymes are known to catalyse the formation of carbanionic intermediates. Flavin-dependent enzymes can also generate radical intermediates. A general strategy to design inhibitors for these enzymes has been to use latent functionalities that can generate during enzymic turnover an electrophilic species, most of the time a so-called Michael acceptor, which eventually alkylates a residue at the active site as shown in Scheme 5.

As we will see from specific examples, because of the principle of microscopic reversibility of enzyme catalysis, both analogues of the substrates and of *the products* of the target enzymes must be considered for design of mechanism-based inhibitors. It is important to realize that this concept applies even when the enzymes catalyse apparently irreversible reactions.

## V. Mechanism-based Inhibitors of PLP-dependent Transaminases

We shall illustrate the principles which led to design mechanism-based inhibitors of PLP-dependent transaminases by discussing the case of 4-aminobutyric acid: 2-oxoglutarate amino transferase (EC 2.6.1.19; GABA-T). In analogy with the mechanism-based inhibition of PLP-dependent aspartate transaminase by serine-O-sulphate, Fowler and John[33] synthesized and tested the higher homologue ethanolamine-*O*-sulphate (EOS) as a potential inhibitor of GABA-T. EOS, an isoster of GABA, proved to be the first efficient mechanism-based inhibitor of GABA-T. The mechanism of inhibition was rationalized as depicted in Scheme 6. The key step is the elimination of a sulphate ion to generate an electrophilic conjugated imine.

EOS has the advantage of being specific for GABA-T but the disadvan-

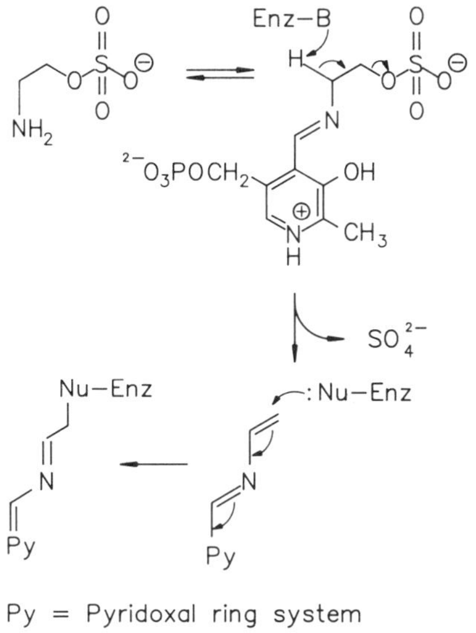

Py = Pyridoxal ring system

SCHEME 6

tage of only poorly penetrating the central nervous system where its therapeutic effect is realized.

Recognizing the potential of EOS, but being aware of its limitation, Metcalf[34] set about designing a potent GABA-T inhibitor that would penetrate the central nervous system. Germane to the proposed starting point of Metcalf's GABA-T inhibitor programme was the enzymic acetylene to allene conversion demonstrated by Bloch. Rationalizing a similar possibility for GABA-T (see Scheme 7, pathway a) Metcalf synthesized γ-acetylenic GABA and showed it to be a potent, mechanism-based inhibitor of the target enzyme.[15]

More detailed biochemical evaluation of the specificity of γ-acetylenic GABA *in vitro* and *in vivo* revealed that the compound was also a mechanism-based inhibitor of glutamate decarboxylase (EC 4.1.1.15; GAD), the enzyme responsible for the synthesis of GABA.[35,36] This unexpected observation was rationalized on the basis of the microscopic reversibility principle of enzyme catalysis. Subsequent exploitation of this perspicacious explanation led to the design of novel potent mechanism-based inhibitors for GABA-T[37-39] and other PLP-dependent enzymes which will be discussed later.

By analogy with the mechanism of inactivation of GABA-T by γ-acetylenic GABA depicted in paths a and b of Scheme 7, it was predicted[40,41]

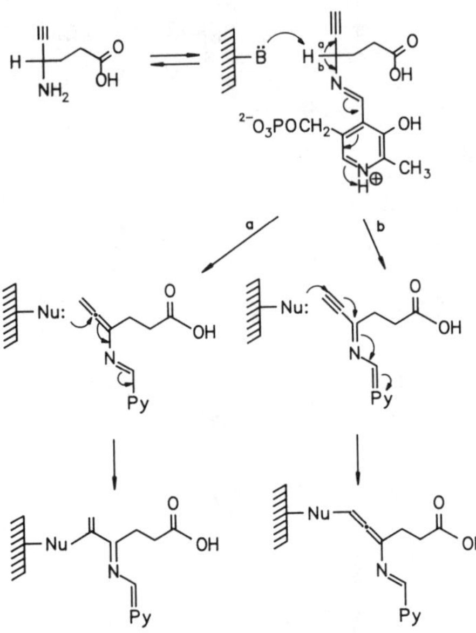

SCHEME 7

that $\gamma$-vinyl GABA would also inactivate GABA-T as shown in Scheme 8, pathway a or alternatively pathway b.

Building on the proposition that transaminases in general and GABA-T in particular could initiate elimination of good leaving groups positioned on the carbon atom $\beta$ to the carbon bearing the amino function of the substrate,

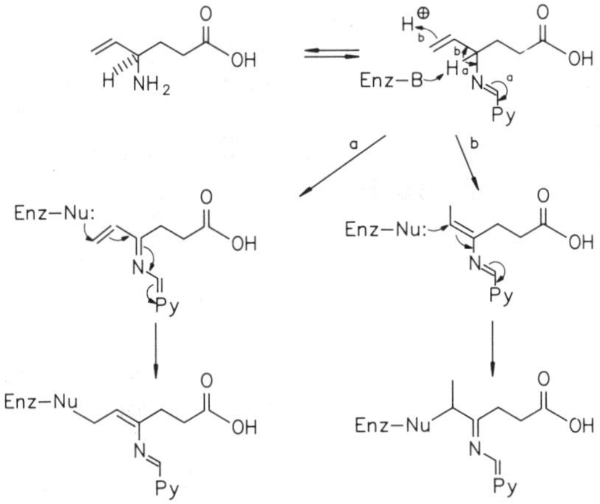

SCHEME 8

TABLE 1

Kinetic constants for the inhibition of GABA-T[a] by GABA derivatives

| Inhibitors | $K_I$ (mM) | $\tau_{50}$ (min) | $t_{1/2}$ (min) at 1 mM |
|---|---|---|---|
| γ-Acetylenic GABA | 0·08 | 6·0 | 6·0 |
| γ-Vinyl GABA[c] | | | 1·0[b] |
| γ-Allenyl GABA[c] | | | 0·5[b] |
| γ-Fluoromethyl GABA | 2·0 | 1·8 | 5·0 |
| γ-Difluoromethyl GABA | 20·0 | 1·8 | 40·0 |
| γ-Trifluoromethyl GABA | | no inhibition | |

[a] Purified from pig brain; the $K_M$ of GABA under the same experimental conditions as those used to study the inhibitors was 4 mM.
[b] No apparent saturation kinetics were observed. [c] $K_I$ and $\tau_{50}$ were not determined.

a series of substrate analogues incorporating a methyl substituent, functionalized by leaving groups, on the amine-bearing carbon atom were synthesized by Silverman and his collaborators[42–44] and by our group.[45] Such analogues and most particularly the fluoromethyl derivatives proved to be potent mechanism-based inhibitors of GABA-T.

A plethora of GABA analogues have been designed and synthesized as mechanism-based inhibitors of GABA-T. Table 1 gives a flavour of the latent reactive groups which have been used and an indication of their potency as measured by the relevant kinetic constants $K_I$ and $\tau_{50}$. Two of these inhibitors, γ-acetylenic GABA (GAG) and γ-vinyl GABA (vigabatrin) were selected for clinical evaluation. Studies with GAG had to be discontinued due to untoward side-effects probably resulting from inhibition of GAD. Vigabatrin is now in phase III clinical studies for the treatment of certain forms of epilepsy which are resistant to currently available therapy.[46] The success of vigabatrin in this respect is certainly due to its specificity.

### VI. Mechanism-based Inhibitors of PLP-dependent Decarboxylases: Analogues of the Substrate

1-Carboxylyases are of major importance in the processing of α-amino acids and the synthesis of amine neurotransmitters and neuromodulators. Selective inhibitors for these enzymes would be expected to be of therapeutic value.

With the exception of S-adenosylmethionine decarboxylase, all mammalian 1-carboxylyases are PLP dependent. As in the case of the PLP-dependent transaminases, the first step in the enzymatic decarboxylation reaction

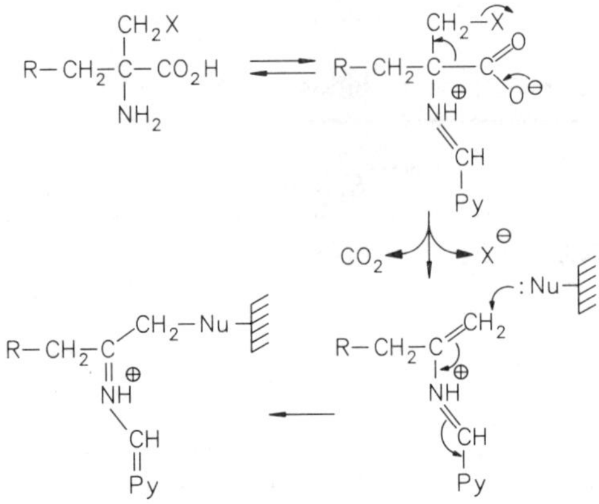

SCHEME 9

is the formation of a Schiff base between the $\alpha$-amino group of the substrate and the aldehyde function of PLP.[47] This is then followed by cleavage of the bond of the carboxylate to the $\alpha$ carbon atom of the substrate which generates a carbanionic intermediate similar to the one formed during the reaction catalysed by transaminase. Understanding the mechanism of decarboxylation led to the proposal and synthesis of a number of $\alpha$-functionalized methyl analogues of amino acids as potential mechanism-based inhibitors of the corresponding 1-carboxylyases.[48] This general approach is rationalized in Scheme 9 for $\alpha$-halogenomethyl amino acids. Interestingly these $\alpha$-amino acid derivatives are deprived of the $\alpha$ hydrogen atom. Consequently they cannot be turned over and, therefore, they cannot be mechanism-based inhibitors of other PLP-dependent enzymes such as transaminases, racemases, etc., which initiate catalysis by abstraction of the $\alpha$ hydrogen atoms of the substrate.

## A. INHIBITORS OF ORNITHINE DECARBOXYLASE
## (EC 4.1.1.17; ODC)

ODC is one of the rate-limiting enzymes in the biosynthesis of putrescine and the polyamines spermidine and spermine. A number of $\alpha$-functionalized methylornithine derivatives were synthesized and tested as inhibitors of ODC[49] (Table 2). Of these, $\alpha$-difluoromethylornithine (DFMO) was found to be the most effective inhibitor. DFMO has been the "workhorse" in polyamine research for the last ten years. It has been useful in establishing a major role for polyamines in cell division and cell differentiation and is proving of considerable value in controlling infections caused by polyamine-sensitive protozoa such as African trypanosomes and *Pneumocystis carinii*.

TABLE 2

Inhibitory properties of $\alpha$-substituted ornithine
analogues towards ODC[a]

| $\alpha$-Substituent | Type of inhibition | $K_I$ (mM) | $\tau_{50}$ (min) |
|---|---|---|---|
| $CH_3$ | Competitive | 0·04 | |
| $CH_2OH$ | No inhibition | | |
| $CH_2OCH_3$ | No inhibition | | |
| $CH_2CN$ | Time-dependent | 8·70 | 29·0 |
| $CH_2Cl$ | Time-dependent | No saturation kinetics[b] | |
| $CH_2F$ | Time-dependent | 0·075 | 1·6 |
| $CHF_2$ | Time-dependent | 0·039 | 3·1 |

[a] Partially purified from a preparation of livers of thioacetamide-treated rats. The $K_M$ of L-ornithine is 0·04 mM.
[b] At 0·1 mM, the half-life of ODC activity is 22 min.

Results of kinetic studies[48,49] with this series of ornithine analogues are consistent with the mode of inactivation proposed in Scheme 9 ($R=(CH_2)_2NH_2$). Thus, the rate of inactivation of ODC by DFMO follows pseudo first-order kinetics with no lag time before the onset of inhibition, is not modified in the presence of dithiothreitol in the incubation medium, and is reduced in the presence of ornithine or putrescine; saturation is observed, and the inhibitory activity resides mainly with the (-) isomer.

Because of the instability and the limited availability of purified ODC, mechanistic studies with ODC inhibitors have been limited so far to DFMO.[23] Radioactive DFMO labelled with [$^{14}$C] at the 5 position was found to bind stoichiometrically to one subunit of ODC, purified from mouse kidney stimulated with androgen, through a covalent bond with Lys 298. In addition, 3 mol of $^{14}CO_2$ are released from 1-[$^{14}$C] DFMO before inhibition of ODC is completed, i.e. the partition ratio of decarboxylation to inhibition is about 3. These results fully support the mechanism depicted in Scheme 9. However, as we shall see in Section X, this scheme could be superseded by an alternate mechanism proposed by Metzler et al..

## B. INHIBITORS OF L-AROMATIC AMINO ACID DECARBOXYLASE (EC 4.1.1.26; AADC)

On the basis of the similarity of the mechanism of action of all PLP-dependent 1-carboxylases, the approach outlined in the previous section for inhibiting ODC is also applicable to the inactivation of other $\alpha$-amino acid decarboxylases. In 1978, we prepared and tested the monofluoromethyl and difluoromethyl analogues of 3,4-dihydroxyphenylalanine (DOPA) and as predicted found them to meet the kinetic criteria for mechanism-based inhibitors of L-aromatic-$\alpha$-amino acid decarboxylase.[48,50,51] However, in contrast to the inhibition of ODC with mono and difluoromethylornithines, the rate of inactivation did not follow first-order kinetics. At the same time, scientists from the Merck laboratories independently synthesized $(S)$-$\alpha$-monofluoromethyl DOPA.[52] Using 1-[$^{14}$C] and [$^3$H] ring-labelled inhibitors they demonstrated elegantly that the inactivation of 1 mol of AADC was accompanied by release of 1 mol of $CO_2$ and irreversible attachment of 1 mol of inhibitor to the enzyme.[22] In addition, as purified AADC can be easily obtained in large quantity, they were able to measure with a fluoride electrode a release of 1·3 mol of fluoride ion for every mol of AADC inactivated.[22] These results indicate that the partition ratio of inhibitor turnover to inhibition is close to 1. They are also in full agreement with the original proposal for the inactivation[48] depicted in Scheme 9 ($R$=3,4-dihydroxyphenyl).

AADC is also known to decarboxylate 5-hydroxytryptophan. The $\alpha$-mono and difluoromethyl analogues of 5-hydroxytryptophan were synthesized, and as expected were found to be mechanism-based inhibitors of AADC.[53,54]

Monofluoromethyl DOPA is the most potent inhibitor of AADC yet reported. It has proven to be a useful pharmacological tool to block monoamine synthesis both centrally and in the periphery.[55] Difluoromethyl DOPA is less potent than the monofluoro analogue, but interestingly it appears to be specific for the enzyme in the periphery.[50]

## C. INHIBITORS OF HISTIDINE DECARBOXYLASE
### (EC 4.1.1.22; HDC)

Similarly, the $(S)$ enantiomer of monofluoromethyl histidine (MFMH) is a potent mechanism-based inhibitor of PLP-dependent mammalian[52,56] and *Morganella morganii* AM-15 HDC. Under optimal conditions a single molecule of MFMH inactivates one enzyme subunit of *Morganella morganii* indicating a partition ratio of 1. For the foetal rat HDC the ratio of $CO_2$ evolved to that of MFMH incorporated into the protein is 2·8.[56] In both cases there is a correlation between the extent of incorporation of labelled inhibitor into the enzyme and the degree of inactivation. The bound inhibitor is not released by dialysis, but is released upon denaturation by heat or urea. In contrast to inhibition of ODC and AADC, where the difluoromethyl analogues are potent inhibitors, difluoromethyl histidine only weakly inactivates mammalian HDC (unpublished). Furthermore, MFMH does not inhibit the pyruvate-dependent HDC from *Lactobacillus*. Similarly, *S*-adenosylmethionine decarboxylase, the only mammalian 1-carboxylase known to be pyruvate-dependent, is not inactivated by the α-difluoromethyl analogue of its substrate.

MFMH is an efficient inhibitor of HDC *in vivo*.[57] It is currently being evaluated in the clinic for the treatment of pathological conditions where excess histamine is believed to be implicated.

## D.  INHIBITORS OF GLUTAMIC ACID DECARBOXYLASE (EC 4.1.1.15; GAD)

Continuing the relationship established for other mechanism-based inhibitors of 1-carboxylyases, α-monofluoromethyl glutamate inhibits in a time-dependent manner both bacterial and mammalian GAD.[52,58] The difluoromethyl analogue, although fairly potent against the bacterial enzyme, is totally inactive against the mammalian GAD (unpublished). GAD is also susceptible to inactivation by α-methyl-*trans*-dehydroglutamic acid.[59] The proposed mechanism of inactivation is shown in Scheme 10.

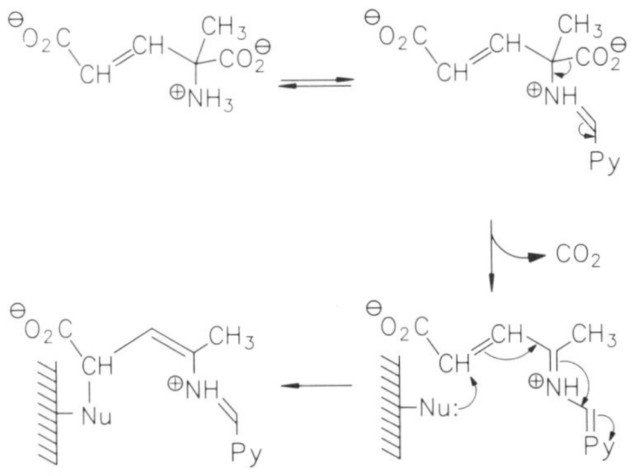

SCHEME 10

Neither the α-fluoromethyl nor the α-methyl-*trans*-dehydro analogues of glutamic acid were active *in vivo*. Consequently their use has been somewhat limited. Fortunately, an alternative strategy, discussed in Section VII, relying on the principle of microscopic reversibility of enzyme catalysis, can be employed to design GAD inhibitors which are active *in vivo*.

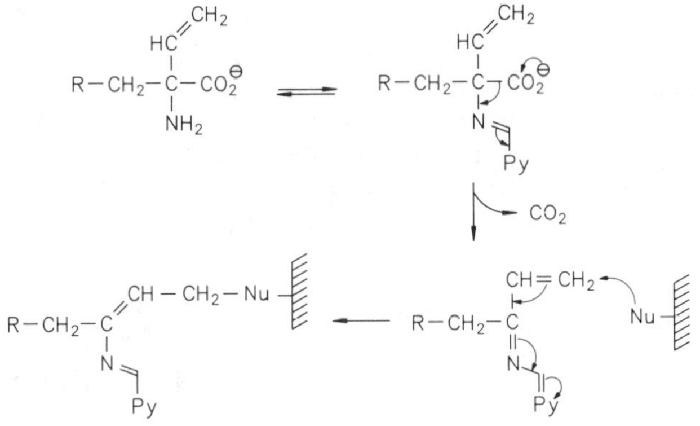

SCHEME 11

## E. OTHER LATENT GROUPS FOR DECARBOXYLASE INHIBITORS

Despite the similarity of their mode of action, there is a striking selectivity of these different $\alpha$-fluoromethyl amino acids for their respective enzymes. For example, high concentrations of MFMH do not inhibit AADC whereas high concentrations of $\alpha$-monofluoromethyl DOPA are inactive against HDC. This selectivity is a reflection of the specificity of the decarboxylase for their respective substrates.

Fluorine is the latent functionality of choice when a leaving group is required in the design of mechanism-based inhibitors.[60,61] Fluorine has a small size and its ease of elimination in $E_1$ reactions compared to the strength of its bond with a carbon atom are important attributes. However, latent groups other than a leaving group can also be used to design mechanism-based inhibitors of decarboxylases. For example, $\alpha$-amino acid analogues incorporating an $\alpha$-acetylenic, vinylic, or allenic substituent usually meet the minimum kinetic criteria for mechanism-based inhibitors of the corresponding decarboxylases. In general, the acetylenic analogue is the most potent of these unsaturated inhibitors and its potency is often in the range of that of the fluoromethyl derivatives. The mode of inactivation of these unsaturated analogues can be rationalized as depicted in Scheme 11 for the ethylenic derivatives.

Fairly detailed analysis of the mechanism of inactivation of AADC by $\alpha$-acetylenic[62,63] and $\alpha$-vinyl DOPA[63] have been published.

# VII. Microscopic Reversibility Principle as an Aid in Inhibitor Design: Analogues of the Products

Initial *in vivo* studies with $\gamma$-acetylenic GABA in mice revealed that the expected inhibition of GABA-T was also accompanied by significant inactivation of GAD.[35] Subsequently, $\gamma$-acetylenic GABA was established *in vitro* to be a mechanism-based inhibitor of GAD.[64-66] A mechanism relying on the microscopic reversibility principle of enzyme catalysis was proposed to rationalize this unexpected finding. As an analogue of the product GABA, $\gamma$-acetylenic GABA enters the active site of GAD and forms a Schiff base with PLP. Then the basic residue whose conjugated acid protonates the carbanionic intermediate in the decarboxylation reaction can catalyse the reverse reaction, i.e. the abstraction of the $\alpha$ hydrogen atom. As indicated in Schemes 12b and 7, the result is the formation in the active site of GAD of the very same intermediate that eventually leads to the inactivation of GABA-T.

This approach has been extended successfully to the inactivation of other $\alpha$-amino acid decarboxylases. Moreover, as the carbanionic intermediate generated from the product analogue is identical to the one formed from the substrate analogue, the same latent groupings can be used to design mechanism-based inhibitors both in the product and substrate series. Thus, the $\alpha$-fluoromethyl and $\alpha$-acetyleneic analogues of putrescine,[67] agmatine,[68] histamines[69] and dopamine[52] are potent mechanism-based inhibitors of ODC, arginine decarboxylase, HDC, and AADC, respectively.

1-Carboxylyases catalyse the decarboxylation of $(S)$ $\alpha$-amino acids with retention of configuration, i.e. the newly added hydrogen atom in the product has the same stereochemistry as the carboxylic group in the $(S)$ $\alpha$-amino acid. The microscopic reversibility principle implies that this very same hydrogen atom must be abstracted in the inactivation of the enzyme by the product analogue. Therefore, the inhibitory activity in the product analogues series should reside with the $(R)$ isomers. This has generally been found to be the case. ODC[70] and bacterial GAD[66] are inactivated by the $(R)$ enantiomers of acetylenic analogues of putrescine and GABA whereas the $(S)$ enantiomers are totally devoid of inhibitory activity. However, there are cases where the inhibition is not as stereospecific. For example, both $(S)$ and $(R)$ isomers of $\alpha$-fluoromethyl dopamine and histamine inhibit AADC and HDC, respectively.[52] An extreme example is mammalian GAD which is solely inactivated by the "wrong" $(S)$ enantiomer of $\gamma$-acetylenic GABA.[66]

Product analogue mechanism-based inhibitors have also been designed for PLP-dependent transaminases. A characteristic of these enzymes is the

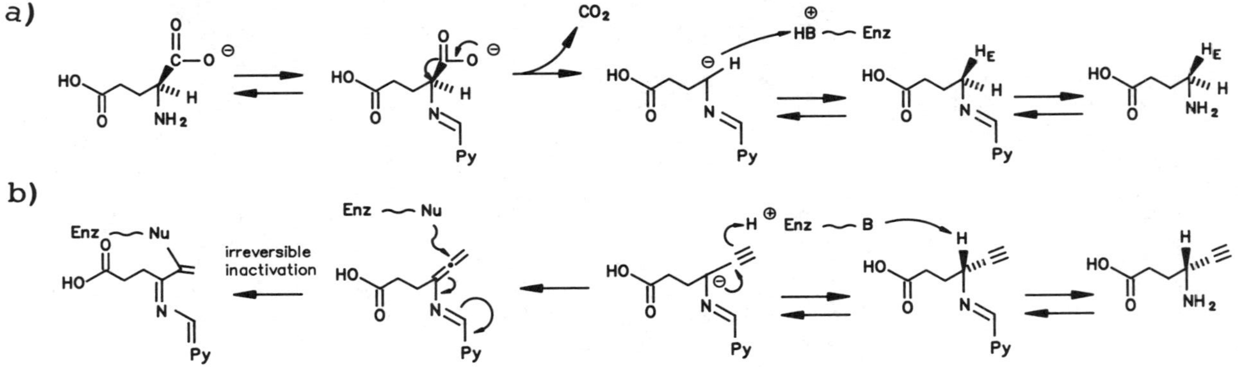

SCHEME 12

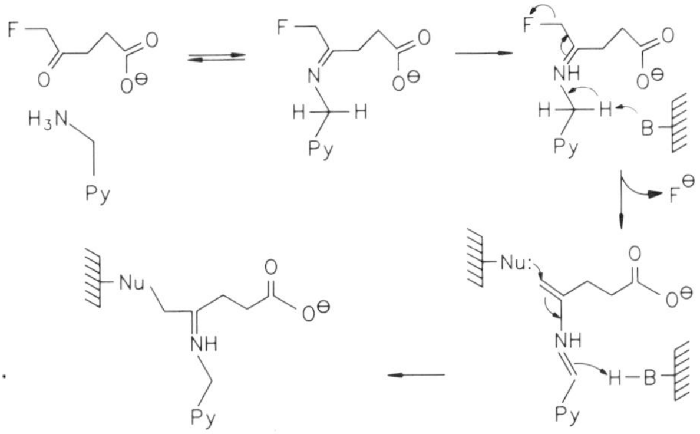

SCHEME 13

conversion of PLP to pyridoxamine during catalysis. PLP has to be regenerated through a transamination reaction usually with an $\alpha$-ketoacid. For example, GABA-T catalyses reversibly the transamination of GABA to succinic semialdehyde and regenerates PLP through a transamination with $\alpha$-ketoglutarate. It was reasoned that suitably substituted analogues of the products succinic semialdehyde or $\alpha$-ketoglutarate should inhibit GABA-T. In practice, 5-fluoro-levulinate was found to inactivate GABA-T.[38] Its proposed mode of action is depicted in Scheme 13.

Clearly, product analogues are a fruitful starting point for the design of mechanism-based inhibitors. However, as indicated by the results obtained for the inhibition of decarboxylases by enantiomers of product analogues, the mode of action of these inhibitors can be complex and does not necessarily follow the pathways predicted by the reversibility principle of enzyme catalysis.

## VIII. Alternative Substrate Analogues as an Approach to Inhibitor Design

Enzymes are often not specific about their substrates. This property can be used to advantage to design mechanism-based inhibitors. In addition to 3,4-dihydroxyphenylalanine, AADC can decarboxylate various ring hydroxylated phenylalanine derivatives. A good correlation was found between the substrate activity of the hydroxyphenylalanine derivatives and the *in vitro* inhibitory potency towards AADC of the corresponding $\alpha$-fluoromethyl analogues.[71] Similarly, GABA-T transaminates $\beta$-alanine as efficiently as GABA and $\gamma$-aminopentanoic acid (homo GABA), the higher homologue of GABA, at one-third the rate of GABA. The monofluoromethyl derivatives of $\beta$-alanine, GABA and homo GABA are all potent inhibitors of GABA-T.[45] Needless to say, there are notable exceptions to the substrate activity/inhibitor potency correlation. Although *trans*-dehydro-GABA is a better substrate than GABA for GABA-T, *trans*-dehydro-$\gamma$-vinyl-GABA proved to be a much weaker inhibitor of GABA-T than $\gamma$-vinyl-GABA.[72] $\alpha$-Monofluoromethyl-*trans*-dehydro-ornithine is another notorious discrepancy.[73,74] It is the most potent mechanism-based inhibitor of ODC reported so far in the ornithine series and yet *trans*-dehydroornithine is not a substrate for ODC.[74]

These examples serve to underscore that mechanism-based inhibitors do *not* have to be good substrates to be powerful and effective inhibitors. In fact, generally mechanism-based inhibitors are extremely poor substrates of the target enzymes. For example, it can be evaluated that under saturating conditions ODC processes DFMO at a rate about 1/5000 that of ornithine.[23] But if each enzyme turnover produces an active species that kills the enzyme, which corresponds to an optimal situation of a partition ratio of 1, then very little processing is required for effective inhibition.

### IX. Importance of the Positioning of the Latent Reactive Group

The right choice and the correct positioning of the latent functionality are crucial for inhibition. Silverman and collaborators[44,75] studied a series of $\beta$-halogeno GABA and dehydro-halogenomethyl GABA derivatives. On the basis of the ability of GABA-T to catalyse elimination reactions, these compounds would be expected to generate reactive intermediates and consequently be potential inhibitors of GABA-T as illustrated in Scheme 14 for the latter derivatives. In fact, although dehydro-halogenomethyl GABA derivatives are weak substrates for GABA-T, time-dependent inhibition was not observed. Only competitive inhibition could be demonstrated. To account for this lack of time-dependent inhibition, Silverman *et al.*[44] offered two possible explanations: (a) that HX elimination did not take place, or (b) elimination occurred but the absence of a suitably positioned nucleophile within bonding distance precluded covalent attachment of the activated moiety. Detection of fluoride ion release during processing of compound, where $X = F$ (pathway a), argues in favour of the second explanation, i.e. the misplaced nucleophile in relation to the active intermediate generated in the active site. Fluoride ion elimination, however, does not occur with every turnover. Determination of glutamate formed from ketoglutarate suggests that transamination is also competing (pathway b) in a 1:4 ratio with elimination. It should be noted that the transamination (pathway b) also generates reactive electrophilic intermediates.

Similarly one might predict that $\beta$-fluoromethylene derivatives of amino acids would be mechanism-based inhibitors of the corresponding decarboxylases as depicted in Scheme 15 for the inactivation of AADC. In fact,

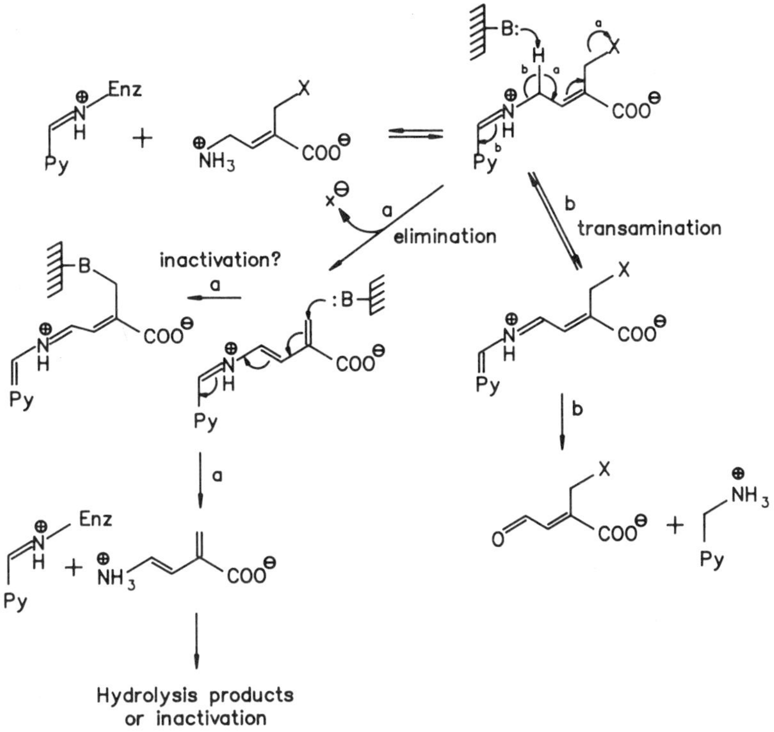

SCHEME 14

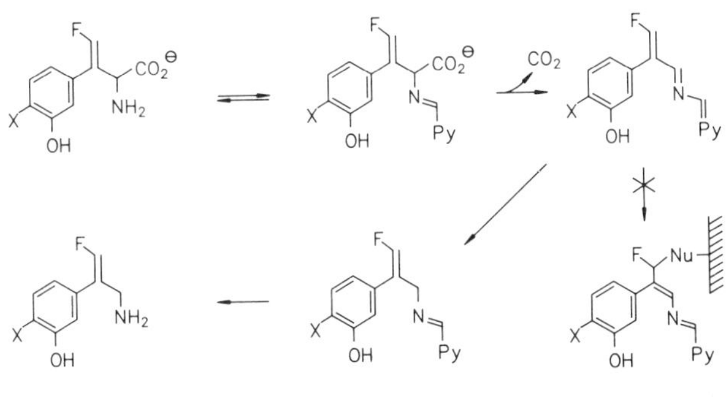

SCHEME 15

ring hydroxylated derivatives of $\beta$-fluoromethylene phenylalanine are excellent substrates and not inhibitors of AADC.[76] The absence of a nucleophilic residue within bonding distance of the electrophilic Michael acceptor could explain the lack of inhibition. Alternatively, the general mechanism proposed in Schemes 9 and 11 for the inactivation of carboxylyases could be incorrect. It is interesting to note that $\beta$-fluoromethylene glutamic acid is neither a substrate nor an inhibitor of GAD.[77]

These examples illustrate one of the difficulties in the design of mechanism-based inhibitors. Generation of activated species in the target enzyme's active site is a necessary, but not sufficient condition for a compound to be a mechanism-based inhibitor.

## X. The Metzler Mechanism of Inactivation of PLP-dependent Enzymes

Metzler and co-workers[78] demonstrated, by determining the structure of the enzyme-inhibitor adduct, that the irreversible inactivation of GAD and aspartate aminotransferase by L-serine-*O*-sulphate does not proceed as postulated through addition of a nucleophilic residue to an enzyme-generated Michael acceptor, but results from the attack of the $\beta$ carbon of an $\alpha$-aminoacrylate intermediate on the internal Schiff base of PLP with a lysine side chain of the protein. The same mechanism has been shown to apply to the inhibition of alanine racemase[79] and GABA-T[42] by $\beta$-fluoroalanine and $\gamma$-fluoromethyl GABA, respectively. Interestingly, in the latter case a partition ratio of 1 was measured for the inhibition which is surprising in view of the fact that the reactive intermediate enamine can readily diffuse from the active site as it is not covalently bound to the enzyme or the cofactor. Preliminary results on the structure of the enzyme–inhibitor adduct indicate that the inhibition of HDC from *Morganella morganii*

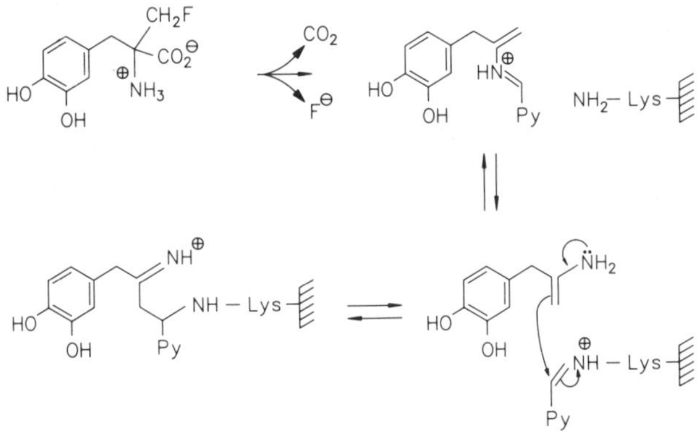

SCHEME 16

AM-15 by $\alpha$-fluoromethyl histidine is also consistent with a similar mechanism.[80] A Metzler type mechanism for the inactivation of AADC depicted in Scheme 16 for $\alpha$-monofluoromethyl DOPA could also explain why $\beta$-halomethylene derivatives of L-*meta*-tyrosine are good substrates and not inhibitors of AADC. Isolation and identification of more enzyme–inhibitor adducts will be required to determine the scope of the Metzler mechanism in the inactivation of PLP-dependent enzymes.

## XI. Mechanism-based Inhibitors of Flavin-dependent Enzymes

Flavin coenzymes are redox coenzymes which have chemical versatility in that they can catalyse two-electron processes as well as one-electron transfers. Flavoproteins catalyse key metabolic oxidation/reduction reactions such as oxidation of alcohols or amines to carbonyl functions, dehydrogenation reactions, oxidation of thiols to disulphides, and hydroxylation of aromatic rings. Because of the chemical versatility of the flavin cofactor, the

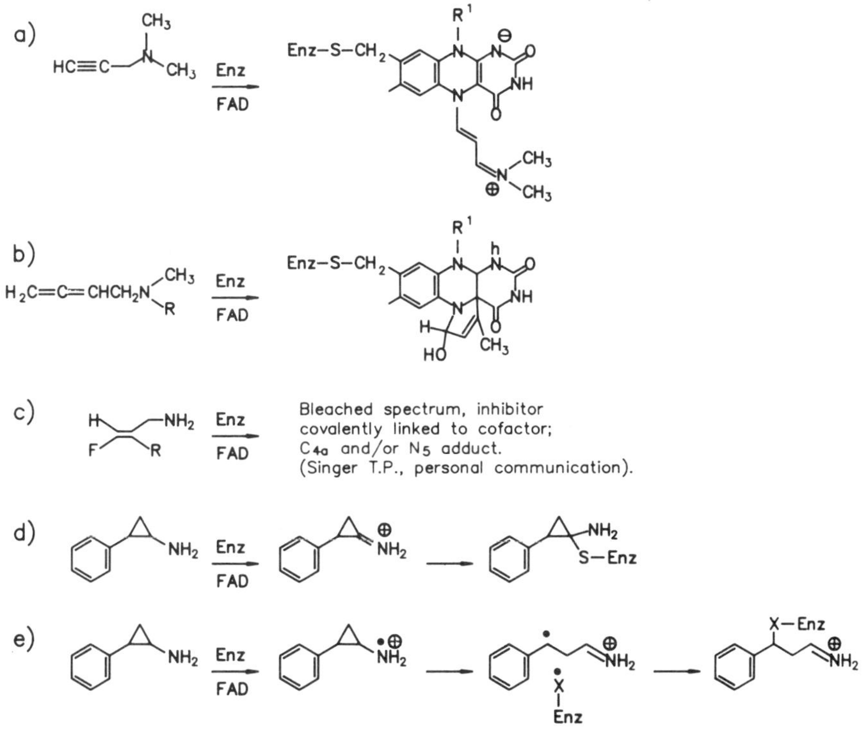

SCHEME 17

exact mechanism by which flavoenzymes transfer electrons is often not known. General strategies, however, have evolved to design mechanism-based inhibitors for these enzymes. These consist of introducing in the substrates of the target enzymes allene, acetylene, cyclopropane, or halomethylene groups in suitable juxtaposition to the functionality that is oxidized. The proposed mechanism of activation of these different latent groups are outlined in Scheme 17. Depending on its state of oxidation the flavin cofactor can act either as a nucleophile or an electrophile. It ensues that the flavin cofactor can compete with nucleophilic residues of the protein to react irreversibly with reactive intermediates generated from the inhibitor.

## A. INHIBITORS OF MONOAMINE OXIDASE (EC 1.4.3.4; MAO)

Monoamine oxidase is á flavin-dependent amine oxidase which uses oxygen as its electron acceptor. It catalyses the oxidative deamination of amines to yield hydrogen peroxide, aldehydes, and ammonia or an amine in the cases of primary amines or secondary and tertiary amine substrates, respectively. MAO plays a key role in the general metabolism of aminergic neurotransmitters. It also deaminates exogenous amines.[81] MAO exists in mammalian mitochondria in two forms designated type A and B and distinguished by their preferred substrates, their distribution, and their sensitivity to inhibitors.[82]

The acetylene-containing inhibitors of MAO, clorgyline, pargyline, and deprenyl have been shown to attach covalently to the N-5 of the flavin cofactor.[83] Surprisingly, the allene-containing inhibitors yield on inactivation different adducts with the cofactor, possibly cyclic N-5,C(4a) compounds.[84] With the halomethylene inhibitors the covalent adduct has not yet been identified.[85,86] The case of the cyclopropane inhibitors is complex. The original mechanism (Scheme 17d) suggested by Paech *et al.*[87] is incorrect. The inhibition results from an initial one-electron oxidation of the amine followed by cyclopropyl ring opening as depicted in Scheme 17e and covalent bond formation with a sulphydryl residue at the active site.

### (1) Selectivity for MAO isoenzymes

The nature of the latent reactive group does not appear to impart any selectivity to the mechanism-based inhibitors. In fact, the irreversible inhibitors clorgyline and deprenyl, whose selectivity led to the current definition of MAO A and B, are propargylamine derivatives and both bind covalently to the N-5 atom of the flavin cofactor. None of the studies aimed at developing an understanding of the substrate–activity relationship regard-

TABLE 3

Substrate and inhibitor selectivity of phenylethylamine analogues

|  |  |  |  |
|---|---|---|---|
| *Substrate* X,Y = H,H |  |  | *Inhibitor* X,Y = F, H |
| $R_1$ | $R_2$ | *Selectivity* | *Selectivity ratio* B/A |
| H | H | B > A | 10·0 |
| H | OCH$_3$ | B | 100·0 |
| OCH$_3$ | OCH$_3$ | B | 100·0 |
| OH | H | A = B | 0·2 |
| OH | OH | A ≥ B | 0·1 |

ing the specificities of substrates and inhibitors for one form of the enzyme or the other have been conclusive so far. With the halomethylene inhibitors, the selectivity seen in the phenethylamine series reflects, to a certain extent, the enzyme specificity for the corresponding substrates[5,88] (Table 3).

This pattern does not hold in other series. For example, MDL 72 638, a fluoroallylamine inhibitor structurally related to the acetylenic inhibitor clorgyline, the most selective MAO A inhibitor known so far, turned out to be an extremely selective inhibitor for the B isoenzyme.[89] This unexpected finding has led to the design of two of the most selective MAO B inhibitors known to date, MDL 72 887[82] and MDL 72 974[90] (see Table 4).

It is important to realize that no substrate binds specifically to MAO A or MAO B. The substrate selectivity is therefore dependent upon the $k_{cat}/K_M$ ratio for each isoenzyme. Similarly, the specificity of mechanism-based inhibitors for each isoenzyme is determined not only by their $K_I$, but also by their $k_{inh}$.

## (2) Site-selective MAO inhibitors

The design of mechanism-based inhibitors provides a rational approach to the selective inhibition of a target enzyme. However, when the target enzyme is widely distributed, it would often be desirable for therapeutic application to restrict the inhibition to specific tissues or organs. As discussed in a later section, this is the case for the inhibition of MAO A. A new approach to achieve site selectivity is to design "dual enzyme-activated" inhibitors. These are mechanism-based inhibitors which are generated by the action of the enzyme preceding the target enzyme in a given metabolic

TABLE 4

MAO B inhibitor selectivity of fluoroallylamines

| | | MAO A[a] | MAO B[a] | Selectivity[b] B/A ($k_{app}$) |
|---|---|---|---|---|
| MDL 72 638 | | 420 nM | 18 nM | 200 |
| MDL 72 887 | | 1500 nM | 6 nM | 10 000 |
| MDL 72 974 | | 680 nM | 4 nM | ~1000 |
| Clorgyline | | 50 nM | 5800 nM | 0·001 |

[a] Concentrations giving 50% inhibition after 15 min preincubation of rat brain mitochondrial MAO using [$^{14}$C]5-hydroxytryptamine and [$^{14}$C]phenylethylamine as substrates for the A and B enzymes, respectively.

[b] Calculated by comparing concentrations that give an equivalent inhibition (approximately 50% in 10 min) of MAO A and MAO B activity in a time-dependent assay.

pathway. This approach takes advantage of the resemblance between the structures of the mechanism-based inhibitor, the substrate of the target enzyme, and consequently the product of the preceding enzyme in the metabolic pathway.

Dopamine is a substrate of both MAO A and MAO B. It is synthesized by the action of AADC on DOPA. AADC is distributed predominantly in aminergic neurones where it is juxtaposed with MAO activity. AADC is not very specific. It can decarboxylate various phenylalanine derivatives incorporating a hydroxyl substituent in the *ortho* or *meta* position of the aromatic ring, as well as a hydroxyl substituent on the $\beta$ carbon atom to form phenethylamine products whose structures are closely related to those of the $\beta$-fluoromethylene phenethylamine MAO inhibitors. With these consider-

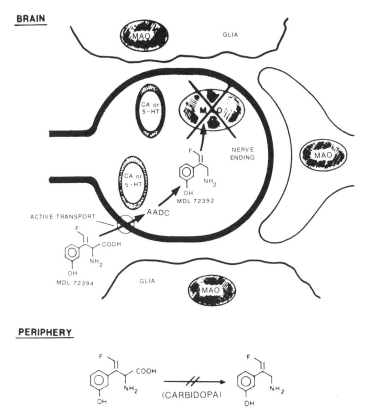

**Fig. 3.** Site-selective inhibition of neuronal monoamine oxidase (MAO) by *E*-β-fluoromethylene-*m*-tyrosine (MDL 72 394).

ations in mind, (*E*)-β-fluoromethylene-*meta*-tyrosine (FMMT) was synthesized as a potential "dual enzyme-activated" inhibitor of MAO A. The *meta*-hydroxy group was expected to confer substrate property to the amino acid for AADC and specificity for MAO A to the corresponding amine. This concept, which is illustrated in Fig. 3, has been validated experimentally in various ways.[91–95]

It is interesting to note that FMMT is a good substrate for AADC and not an inhibitor as is the close analogue α-fluoromethyl-*meta*-tyrosine. As depicted in Schemes 15 and 16, both tyrosine analogues form similar "activated" Michael acceptor intermediates when they are turned over by AADC. As previously discussed, the lack of AADC inhibition by FMMT provides further, albeit circumstantial, evidence supporting the Metzler mechanism for the inactivation of amino acid decarboxylases by α-fluoromethyl amino acids.

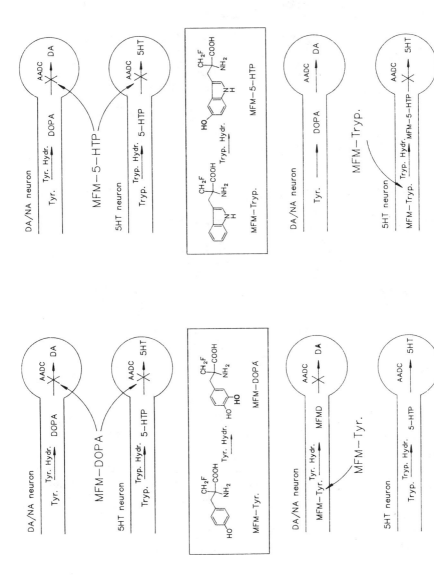

**Fig. 4.** Site-selective inhibition of AADC by α-monofluoromethyl tyrosine (MFM-Tyr) in catecholaminergic neurons and by α-monofluoromethyl tryptophan (MFM-Tryp) in serotonergic neurons. Tyr., tyrosine; Tryp., tryptophan; Hydr., hydroxylase.

The concept of "dual enzyme-activated" inhibitors has also been used successfully to design $\alpha$-monofluoromethyl tyrosine (MFM-Tyr)[93,96] and $\alpha$-monofluoromethyl tryptophan (MFM-Tryp)[53,54] as specific inhibitors of catecholamine biosynthesis and indolamine biosynthesis, respectively. Catecholamine and indolamine synthesis requires the action of AADC on DOPA and 5-hydroxytryptophan (5-HTP). Mechanism-based inhibitors of AADC block both biosynthetic pathways. However, DOPA and 5-HTP are synthesized from tyrosine and tryptophan through the selective action of tyrosine hydroxylase and tryptophan hydroxylase, which are localized in catecholamine and indolamine neurones, respectively. MFM-Tyr and MFM-Tyrp have been found to be acted upon by tyrosine hydroxylase and tryptophan hydroxylase, respectively, to generate the AADC inhibitors selectively in catecholamine or indolamine neurones. The concept is summarized in Fig. 4.

## B. INHIBITORS OF POLYAMINE OXIDASE (PAO)

The flavin-dependent polyamine oxidase catalyses the oxidative cleavage of $N^1$-acetylspermidine and $N^1$ acetyl-spermine to putrescine and spermidine, respectively, with concomitant formation of acetamidopropionaldehyde. PAO may be of importance in regulating the re-utilization of polyamines and consequently the polyamine concentrations in various tissues. The mechanism of action of PAO has many similarities to that of MAO. As MAO is susceptible to irreversible inactivation by acetylenic and allenic substrate analogues, unsaturated analogues of putrescine were synthesized as potential mechanism-based inhibitors of PAO.[97] As indicated in Table 5, the allenic derivatives of putrescine were found on the basis of kinetic criteria only to be extremely potent mechanism-based inactivators of

TABLE 5

Inhibitors of polyamine oxidase (PAO)

| | $R_1NH(CH_2)_4NHR_2$ | | | |
|---|---|---|---|---|
| Compound | $R_1$ | $R_2$ | $K_I(\mu M)$ | $\tau_{50}(\min)$ |
| 1 | $CH_2 = C = CHCH_2$ | H | 0·7 | 1 |
| 2 | $CH_2 = C = CHCH_2$ | $CH_2$ | 0·3 | 0·5 |
| 3 | $CH_2 = C = CHCH_2$ | $CH_2 = C = CHCH_2$ | 0·09 | 2·2 |
| 4 | $CH_2 = CHCH_2$ | H | $t_{1/2} = 35$ min at 50 $\mu M$ | |
| 5 | $CH_2 = CHCH_2$ | $CH_3$ | $t_{1/2} = 12$ min at 12 $\mu M$ | |
| 6 | $CH\ CCH_2$ | H | 425 | 260 |
| 7 | $CH\ CCH_2$ | $CH_3$ | 310 | 100 |

PAO, while the allylic and acetylenic derivatives are considerably less active inhibitors.[97] This difference in potency contrasts with that observed with corresponding MAO inhibitors where acetylenic derivatives display the highest activity. *In vivo* the allenic inhibitors show exquisite potency and selectivity for PAO inhibition.[98]

### XII. Inhibitors of Copper-dependent Amine Oxidases

The so-called $Cu^{++}$-dependent amine oxidases constitute a family of enzymes of considerable biological interest. They catalyse the oxidative deamination of mono-, di-, and polyamines with the formation of stoichiometric amounts of aldehyde, hydrogen peroxide, and ammonia. Members include plasma amine oxidase; an enzyme rich in vascular tissue called by several names including benzylamine oxidase, semicarbazide-sensitive amine oxidase, clorgyline-resistant or pargyline-resistant amine oxidase; diamine oxidase; and lysyl oxidase. Plasma amine oxidase[99] and lysyl oxidase[100] have recently been shown to use covalently-bound pyrroloquinoline quinone as the prosthetic group. Plasma amine oxidase is irreversibly inhibited by 2-chloroallylamine[101] and glycine esters[102] where the alcohol part of the ester is a good leaving group. The design of mechanism-based inhibitors for flavin and $Cu^{++}$-dependent amine oxidases often relies on the use of similar latent groups. For example, propargylamine is an irreversible inhibitor of plasma amine oxidase;[101] (*E*)-2-phenyl-3-chloroallylamine is a potent mechanism-based inhibitor of vascular amine oxidase.[5] *In vitro* this inhibitor has nanomolar affinity for benzylamine oxidase of rat aorta but requires 65 000- and 1500-fold higher concentration to interact with MAO A and MAO B, respectively. The functional role of this benzylamine oxidase is unknown and such a potent and selective inhibitor may aid in determining the physiological significance of this widely distributed enzyme.

### XIII. Inhibitors of Dopamine-$\beta$-hydroxylase (EC 1.14.17.1; D$\beta$H)

D$\beta$H is a copper-dependent monooxygenase which catalyses the conversion of dopamine to norepinephrine. Until recently the only inhibitors available for this enzyme were the rather non-selective copper chelators such as fusaric acid. In 1984, a number of mechanism-based inhibitors were described including $\beta$-chlorophenethylamine,[103] 4-hydroxybenzylcyanide,[104] 2-halo-3-(*p*-hydroxy-phenyl)-1-propenes,[105] 1-phenyl-1-propyne,[106] and 2-phenylallylamine.[107] None of these compounds were particularly potent. Contrary to previous belief,[108] certain heteroaromatic amines were

TABLE 6

Constants of dopamine $\beta$-hydroxylase inhibitors

| Compound | $K_I$ ($\mu$M) | $k_{cat}$ (min$^{-1}$) | $k_{cat}/K_I$ (M$^{-1}$min$^{-1}$) |
|---|---|---|---|
| thienyl-CH=CH-CH$_2$NH$_2$ | 35 | 0·124 | 3540 |
| thienyl-CH=C(CH$_3$)-CH(NH$_2$) | 360 | 0·058 | 161 |
| thienyl-CH=CH-CH$_2$NHCH$_3$ | 2150 | 0·098 | 44 |
| 3-thienyl-CH=CH-CH$_2$NH$_2$ | 570 | 0·147 | 257 |
| furyl-CH=CH-CH$_2$NH$_2$ | 410 | 0·130 | 317 |
| 3-furyl-CH=CH-CH$_2$NH$_2$ | 230 | 0·077 | 334 |
| phenyl-C≡C-CH$_2$NH$_2$ | 21 | 0·07 | 3333 |
| phenyl-CH=CH-CH$_2$NH$_2$ | 13 000 | 0·04 | 3 |
| phenyl-C≡C-CH$_3$ | ~20 000 | 0·06 | 6 |

found by scientists at the Merrell Dow laboratories to be substrates of D$\beta$H. Applying the rationale of inhibition of D$\beta$H produced by 2-phenylallylamine to the new heterocyclic amine substrates, a series of thienylallylamines were synthesized and shown to be very potent mechanism-based inhibitors of D$\beta$H (see Table 6).[109]

Preliminary studies (Hornsberger, unpublished observations) suggest that the mechanism of inactivation of D$\beta$H by the thienylallylamines follows the same steps as those suggested by Padgette and May[110] for the inhibition by phenylallylamine. The proposed mechanism is shown in Scheme 18.

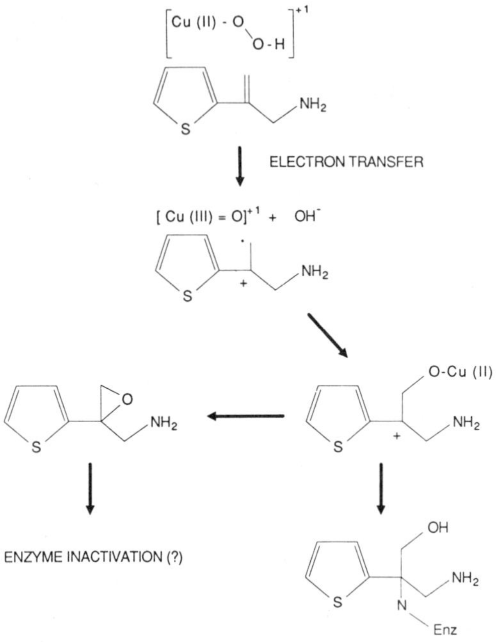

SCHEME 18

## XIV. Mechanism-based Inhibitors that do not involve Chemically-reactive Intermediates

As we outlined in the earlier part of this chapter, mechanism-based inhibition is not necessarily synonymous with the formation of reactive intermediates which eventually react covalently with the protein. Chemically inert dead-end complexes can also be formed as a result of enzyme processing of the inhibitors. These complexes can be covalently bound to the protein or dissociate so slowly from the active site that for all practical purposes they can be considered as irreversible inhibitors. Three illustrative examples of such inhibitors will be discussed briefly.

### A. INHIBITION OF THYMIDYLATE SYNTHASE (EC 2.1.1.45)

5-Fluorouracil (5-FU) is an antimitotic agent used extensively in cancer chemotherapy. *In vivo*, 5-FU is transformed to the active metabolite (5-fluoro-2'-deoxyuridylic acid; FdUMP) which is a potent inhibitor of thymidylate synthase, the enzyme catalysing the only *de novo* synthesis of 2'-deoxythymidylic acid from 2'-deoxyuridylic acid (dUMP). The

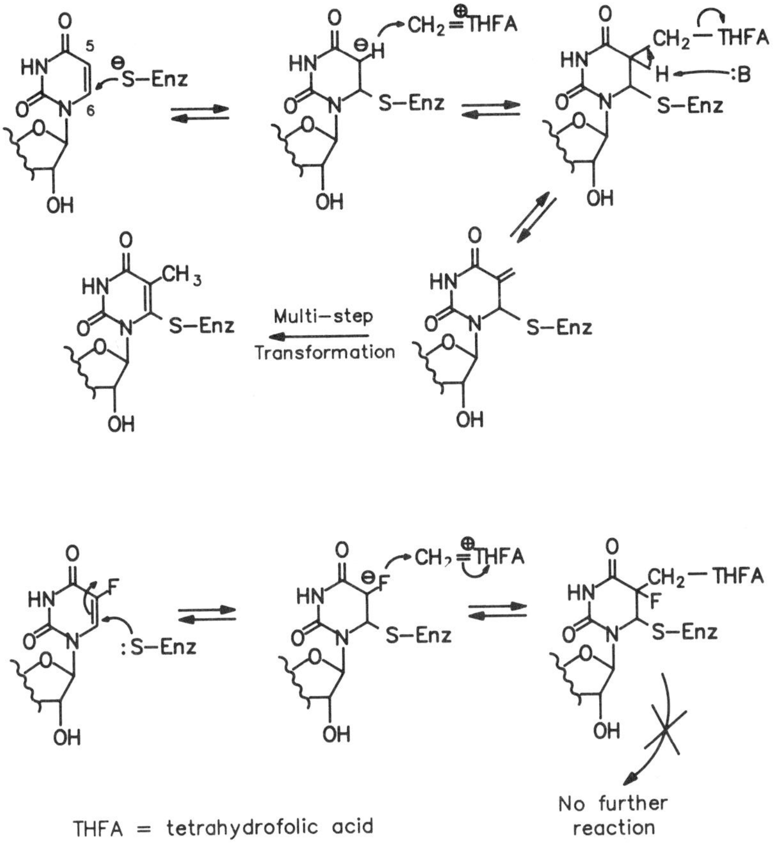

THFA = tetrahydrofolic acid

No further reaction

SCHEME 19

mechanism of action of thymidylate synthase is now well understood and the mechanism of its inactivation by FdUMP has been fully elucidated and characterized.[111] These mechanisms are summarized in Scheme 19.

The reaction is initiated by the nucleophilic attack of a cysteine-198 to the 6 position of dUMP. The resulting carbanion in the 5 position then reacts with methylene tetrahydrofolate to give a ternary intermediate covalently bound to the protein through the cystein residue. The next step requires the abstraction of the hydrogen atom at the 5 position of the ternary complex. FdUMP behaves as a quasi-substrate for thymidylate synthase and forms a ternary complex which has in the 5 position a fluorine atom instead of a hydrogen atom. The carbon fluorine bond is a much stronger bond than the carbon hydrogen bond. The enzyme is unable to cleave the C—F bond and the reaction stops in a dead-end position in mid-catalytic cycle. The enzyme can only be reactivated by the reversion of the first steps leading to the formation of the ternary complex which under physiological conditions has a half-life of between 10 and 14 h.

## B. INHIBITION OF XANTHINE OXIDASE (EC 1.2.3.2; XO)

Xanthine oxidase catalyses the oxidation of hypoxanthine to xanthine and then to uric acid (Scheme 20). XO contains one molecule of flavin adenine dinucleotide and one molecule of molybdenum ($Mo^{VI}$) ion per subunit. XO can reduce $O_2$ in one-electron steps, i.e. XO is capable of superoxide anion formation.[112] XO is believed to be implicated in the generation of peroxide and oxygen radicals that cause tissue damage upon reperfusion of ischaemic tissues.[113] Allopurinol, an isomer of hypoxanthine, is a potent inhibitor of XO. It is the drug of choice for the treatment of gout, a condition characterized by an excess of uric acid. As indicated in Scheme 21, it has been established[114] that XO catalyses a 2-electron oxidation of allopurinol at $C_2$ to yield alloxanthine. As a result, the $Mo^{VI}$ is reduced to $Mo^{IV}$ which then forms a tight complex with alloxanthine. This non-covalent complex dissociates slowly with a $t_{1/2}$ of about 300 min and accounts for the therapeutic action of allopurinol.

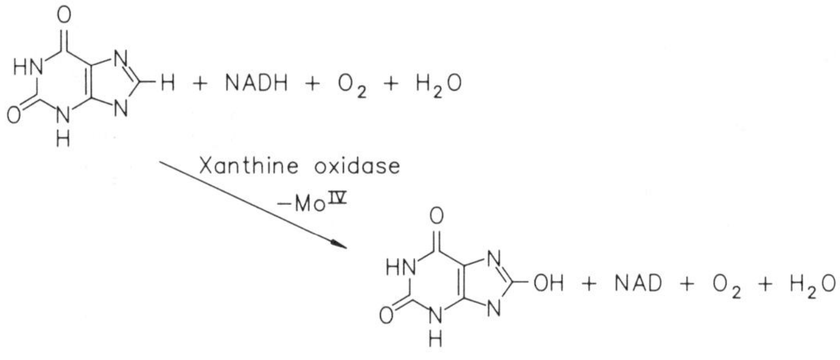

SCHEME 20

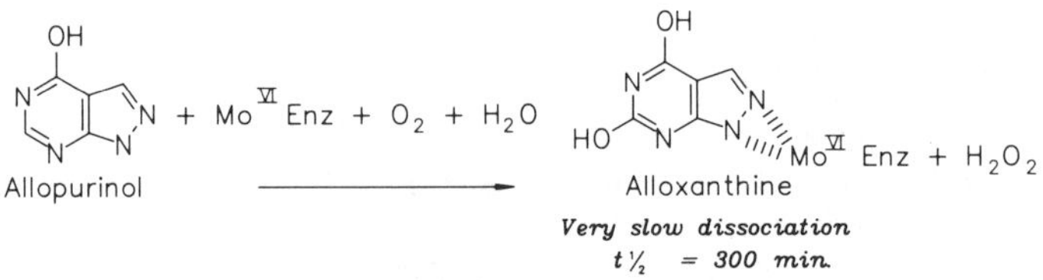

SCHEME 21

## C. INHIBITION OF GABA-T BY GABACULINE

Gabaculine, a natural product isolated from *Streptomyces toyacaenis*, is a potent mechanism-based inhibitor of GABA-T.[115] Although gabaculine shows structural similarity to $\gamma$-vinyl GABA, the mechanism of action of these two inhibitors are totally different. As depicted in Scheme 22, the imine formed upon transamination of gabaculine isomerizes rapidly and irreversibly to a *m*-anthranilic acid derivative which remains tightly bound to the enzyme's active site. As expected, double bond isomers of gabaculine also inhibit GABA-T by yielding the same *m*-anthranilic acid adduct.[116]

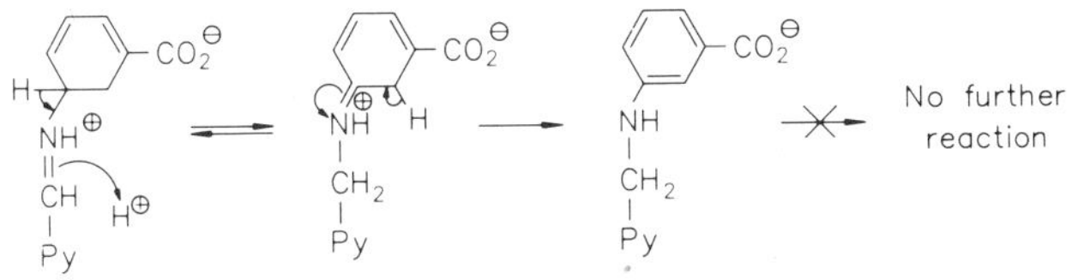

SCHEME 22

This new mechanism of inhibition which relies on an irreversible aromatization of an intermediate along the enzymic reaction pathway was exploited to design other mechanism-based inhibitors. Of particular note is the exquisitely potent inactivation of GABA-T by (*S*)-4-amino-4,5-dihydro-2-thiophene carboxylic acid and (*S*)-4-amino-4,5-dihydro-2-furan carboxylic acid. Mechanistic studies suggested that GABA-T catalyses the formation of thiophene and furan adducts with pyridoxamine, respectively.[117]

## XV. Additional Examples of Mechanism-based Inhibitors

In preceding sections we have attempted by discussing the inactivation of selected target enzymes to give a flavour of the variety of chemical groups that can be exploited in mechanism-based inhibitor design. Other important classes of enzymes are also susceptible to mechanism-based inhibition. For example, hepatic microsomal cytochrome $P_{450}$-dependent enzymes of broad substrate specificity that catalyse the hydroxylation of lipophilic exogenous compounds are inhibited by epoxides resulting from the addition of oxygen across unsaturated carbon—carbon bonds.[118] Inhibition of certain of these enzymes may be of major importance in the metabolic inactivation (or activation) of a large number of xenobiotics. Such "drug" interactions should always be borne in mind when designing inhibitors. Examples do exist where this epoxidation has been exploited beneficially. The inhibition of aromatase, the enzyme responsible for the formation of estrogen from testosterone, is such an example.[119]

Also of note, since no precedent existed, is mechanism-based inhibition of the NADPH-dependent enzyme testosterone $5\alpha$-reductase by $(5\alpha$-$20$-$R$)-4-diazo-21-hydroxy-20-methylpregnan-3-one (MDL 18 341) which has a $K_I$ of $3\cdot5 \times 10^{-8}$ M.[120]

More recently, mechanism-based inhibitors of proteases have been the object of intensive research, although problems of the inherent stability and chemical reactivity of many of the inhibitors has limited this development and has led to the consensus that transition state analogues may be the preferred approach.[121,122]

### XVI. Usefulness of Mechanism-based Enzyme Inhibitors

The inherent specificity, stoichiometry, and irreversible nature of mechanism-based inhibitors has made these compounds valuable tools for biologists. Representative examples of such uses are listed below.

### A. IN BIOCHEMICAL RESEARCH

#### (1) Enzyme turnover

The irreversible nature of the attachment of mechanism-based inhibitors has been exploited to measure enzyme turnover. Specific examples are monoamine oxidase,[123] GABA-transaminase,[21] glutamic acid decarboxylase,[21] aromatic L-amino acid decarboxylase,[124] and ornithine decarboxylase.[125]

#### (2) Titration of enzyme concentration

In many cases the stoichiometry and irreversibility of enzyme inhibitor interaction allows enzyme concentration to be determined accurately. This has been done with [$^3$H] pargyline for brain[126] and liver MAO[127,128] and for ODC,[125] AADC, and GABA-T.[129] The tritium-labelled inhibitors of these enzymes are commercially available. Recently, the fact that $\gamma$-acetylenic GABA inhibits GAD *in vivo* has been exploited to block selectively this enzyme activity in the retina.[129] Radiolabelled $\gamma$-acetylenic GABA administered to animals previously treated with the selective GABA-T inhibitors $\gamma$-vinyl GABA or gabaculine is one obvious strategy for mapping GAD-positive neurons in the central nervous system.

### (3) Autoradiographic studies for enzyme localization

High specific activity tritium-labelled inhibitors of GABA-T and AADC could be used to localize and quantify enzyme in, for example, the central nervous system. The new technique of quantitative computer analysis of autoradiograms is ideally suited to this type of research. Recently this concept has radically been advanced by the use of positron emission tomography in human subjects using selective $C^{11}$-labelled mechanism-based inhibitors of MAO.[130]

### (4) Identification of peptide residues in the enzyme active site

Since alkylation of an amino acid residue in the enzyme's active site is frequently the end result of mechanism-based inhibition, a protein digest of the enzyme-labelled inhibitor adduct can yield useful information on the amino acid sequence around the labelled amino acid. Initial studies with [$^3$H]pargyline in the presence of selective non-labelled inhibitors of either MAO type A or type B has identified a *penta*-peptide common to both enzyme forms.[131,132] Although these two enzymes are distinct proteins (at least antigenically)[133,134] they appear to share common peptides in parts of the active site involved in irreversible binding to inhibitors. Selectivity for these irreversible inhibitors is thus probably determined by the affinities for the active site and not for the site of alkylation.

### (5) Understanding of catalytic mechanism

Mechanism-based inhibitors have proved to be of value to study the catalytic role of the two subunits in GABA-T. Churchich and Moses[135] and Lippert *et al.*[117] have demonstrated elegantly that only one subunit need be inactivated by gabaculine or 4-amino-4,5-dihydro-2-thiophene (or 2-furan) carboxylic acid for enzyme activity to be lost. Similarly, Burnett *et al.*[136] were able to show that incorporation of one [$^{14}$C]propargylglycine molecule per dimer would inactivate, by 97%, pig heart L-alanine transaminase. This half-site reactivity has also been demonstrated for methionine-$\gamma$-lyase and $\gamma$-cystathionase.[18]

The stereochemistry of catalysis has also been elaborated by studing the kinetics of enzyme inactivation with mechanism-based inhibitors that contain a deuterium or tritium-label and/or that have been separated into their pure stereoisomers. For example, results obtained with pure GAD and the stereoisomers of $\gamma$-acetylenic GABA[21] suggest that decarboxylation of glutamate takes place with retention of configuration.

## B. IN PHARMACOLOGICAL RESEARCH

### (1) For measuring turnover of neurotransmitters

A number of methods have evolved using mechanism-based inhibitors for measuring the turnover of amino acid or amine neurotransmitters. Of note are several studies using the rate of accumulation of GABA in various parts of the central nervous system following systemic administration of GABA-T inhibitors such as gabaculine, γ-vinyl GABA, and γ-acetylenic GABA;[137–139] accumulation of 5-hydroxytryptamine and disappearance of its deaminated metabolite 5-hydroxyindole-acetic acid following MAO inhibition[140] or, more recently, measurement of 5-HT, noradrenaline and dopamine turnover in a single brain sample following inhibition of AADC with α-monofluoromethyl DOPA (MFMD).[141]

### (2) Irreversible inhibition of enzymes after micro-injection into specific neuronal tracts

Using micro-injections of γ-vinyl GABA into the substantia nigra of the rat, Gale and Iadorola[138] were able to establish a key role for GABA in this nucleus and this allowed them to conclude that a major control of the motor pathways involved in seizure activity involved transmission via a nigral GABA inhibitory pathway. Recently, we have micro-injected MFMD into cell bodies of the nigrostriatal dopamine pathway and followed the transport of inactive AADC down the axon to the nerve terminals.[92] A measure of axonal flow of various enzymes can be obtained in this way.

### (3) In polyamine research

The significance of DFMO to biomedical research on the role of polyamines was rapidly appreciated and shortly after the initial publication in 1978 by Metcalf and his co-workers[49] key papers appeared showing inhibition of growth of tumour cells in culture[142] and in animals.[143] In 1980 Bacchi *et al.*[144] observed dramatic inhibition of *Trypanosoma brucei brucei* and this led to evaluation of DFMO in a wide variety of experimental parasite infections[145] and subsequent clinical success in treating African sleeping sickness and *Pneumocystis carinii*, an opportunistic protozoan infection frequently associated with acquired immune deficiency syndrome.[146,147] A consequence of the explosive growth of understanding of the significance of polyamine metabolism and its inhibition in therapeutics[147] and the pivotal role played by mechanism-based inhibitors is the *raison d'être* for a recent book on the subject.[148]

## C. USE OF MECHANISM-BASED INHIBITORS
## IN THERAPEUTICS

No review of mechanism-based enzyme inhibitors would be complete without reference to the past, present, and future potential of this class of enzyme inhibitor as therapeutic agents. This subject was reviewed comprehensively by Sjoerdsma in 1981.[149] In many ways their inherent specificity, coupled with the latent reactivity, make them potentially ideal drugs. Reference to Table 7 gives an idea of the diverse therapeutic categories in which mechanism-based enzyme inhibitors have had or will have a major impact. We find it illuminating that many frequently-used and relatively non-toxic drugs were found to be mechanism-based inhibitors long after their therapeutic value had been clearly established; the $\beta$-lactam antibiotics being a classic example. More recently it has become obvious that mechanism-based inhibitors that were rationally designed have been confirmed as being therapeutically effective; DFMO and $\gamma$-vinyl GABA are two such examples. In the initial phases of clinical evaluation are many new mechanism-based inhibitors such as the site selective-MAO inhibitors, which were designed with additional properties to improve their therapeutic potential. We will no doubt see the fruits of "rational" inhibitor design blossom in the next few years.

TABLE 7

Examples of therapeutic categories in which mechanism-based inhibitors are used

| | | |
|---|---|---|
| Antibiotics | Penicillin, cephalosporin, clavulinic acid | Established drugs subsequently found to be mechanism-based inhibitors |
| Anticancer | Fluorouracil, testolactone | |
| Antidepressant | MAO A inhibitors, e.g. tranylcypromine MAO B inhibitors e.g. 1-deprenyl | |
| Antiepileptic | $\gamma$-Vinyl GABA (vigabatrin) | Effective therapeutic agents, designed as mechanism-based inhibitors |
| Antiparasitic | Difluoromethylornithine (eflornithine) | |
| Antidepressant | $(E)$-$\beta$-Fluoromethylene $m$-tyrosine (MDL 72 394) | Designed to be a site-selective mechanism-based inhibitor |
| Antitumoural (estrogen-dependent tumours) | 10-(2-propynyl)estr-4-ene-3,17-dione (MDL 18 962) | Designed as a mechanism-based inhibitor of aromatase |

## XVII. Conclusions

In this chapter we have tried to show the rational (but have frequently cited the irrational) development of this valuable class of mechanism-based enzyme inhibitor. Conceptually, the basis for inactivating enzymes using their own catalytic mechanism is an attractive one. Having been personally involved in the design, synthesis and evaluation of several of these inhibitors, some of which show considerable therapeutic promise, we are convinced that future developments in this area will lead to many more inhibitors of biologically important enzymes and that such development will continue to be of help in unravelling some of the mysteries of complex biological systems.

## REFERENCES

1. Walsh, C. T. (1984). Suicide substrates, mechanism-based enzyme inactivators: Recent developments. *Ann. Rev. Biochem.* **53**, 493–535.
2. Abeles, R. H. & Maycock, A. L. (1976). Suicide enzyme inactivators. *Acc. Chem. Res.* **9**, 313–319.
3. Walsh, C. (1983) Suicide substrates: Mechanism-based enzyme inactivators with therapeutic potential. *Trends Biochem. Sci.* **8**, 254–257.
4. Williams, C. H. (ed.) (1986). Suicide substrates. *Biochem. Soc. Trans.* **14**, 397–413.
5. Palfreyman, M. G., McDonald, I. A., Bey, P., Danzin, C., Zreika, M., Lyles, G. A. & Fozard, J. R. (1986). The rational design of suicide substrates of amine oxidases. *Biochem. Soc. Trans.* **14**, 410–413.
6. *Penguin English Dictionary* (1969). Penguin, Harmondsworth.
7. Rando, R. R. (1974). Chemistry and enzymology of $k_{cat}$ inhibitors. *Science* **185**, 320–324.
8. Seiler, N., Jung, M. J. & Koch-Weser, J. (eds) (1978). *Enzyme-activated Irreversible Inhibitors*. Elsevier/North Holland, Amsterdam.
9. Silverman, R. B. & Hoffman, S. J. (1984). The organic chemistry of mechanism-based inhibition: A chemical approach to drug design. *Med. Res. Rev.* **4**, 415–447.
10. Rando, R. R. (1984). Mechanism-based enzyme inactivators. *Pharmac. Rev.* **36**, 111–142.
11. Kalman, T. I. (ed.) (1979). Drug action and design: Mechanism-based enzyme inhibitors. In *Developments in Biochemistry*, Vol. 6. Elsevier/North Holland, New York.
12. Wolfenden, T. (1976). Transition state analog inhibitors and enzyme catalysis. *Ann. Rev. Biophys. Bioeng.* **5**, 271–306.
13. Brodbeck, U. (1980). On the design of transition state analog enzyme inhibitors and their future in medicinal and agricultural chemistry. In *Enzyme Inhibitors* (Brodbeck, U., ed.), pp. 3–17. Verlag Chemie, Weinheim.
14. Kitz, R. & Wilson, I. B. (1962). Esters of methanesulfonic acid as irreversible inhibitors of acetylcholinesterase. *J. Biol. Chem.* **237**, 3245–3249.

15. Jung, M. J. & Metcalf, B. W. (1975). Catalytic inhibition of $\gamma$-aminobutyric acid:$\alpha$-ketoglutarate transaminase of bacterial origin by 4-aminohex-5-ynoic acid, a substrate analogue. *Biochem. Biophys. Res. Commun.* **67**, 301–306.

16. Waley, S. G. (1980). Kinetics of suicide substrates. *Biochem. J.* **185**, 771–773.

17. Tansumami, J., Yago, N. & Hosoe, M. (1981). Kinetics of suicide substrates. Steady state treatments and computer-aided exact solutions. *Biochim. Biophys. Acta* **662**, 226–235.

18. Johnston, M., Jankowski, D., Marcotte, P., Tanaka, H., Esaki, N., Soda, K. & Walsh, C. (1979). Suicide inactivation of bacterial cystathionine gamma-synthetase and methionine gamma-lyase during processing of L-propargylglycine. *Biochemistry* **18**, 4690–4701.

19. Cromartie, T. H. & Walsh, C. (1975). Mechanistic studies on the rat kidney flavoenzyme-L-alpha-hydroxy acid oxidase. *Biochemistry* **14**, 3482–3489.

20. Endo, K., Helmkamp, M. & Bloch, K. (1970). Mode of inhibition of $\beta$-hydroxydecanoyl thio ester dehydrase by 3-decynoyl-$N$-acetylcysteamine. *J. Biol. Chem.* **245**, 4293–4296.

21. Lippert, B., Jung, M. J. & Metcalf, B. W. (1980). Biochemical consequences of reactions catalyzed by GAD and GABA-T. *Brain Res. Bull.* **5** (Suppl. 2), 375–379.

22. Maycock, A. L., Aster, S. D. & Patchett, A. A. (1980). Inactivation of 3-(3,4-dihydroxyphenyl)alanine decarboxylase by 2-(fluoromethyl)-3-(3,4-dihydroxyphenyl)alanine. *Biochemistry* **19**, 709–718.

23. Pegg, A. E., McGovern, K. A. & Wiest, L. (1987). Decarboxylation of $\alpha$-difluoromethylornithine by ornithine decarboxylase. *Biochem. J.* **241**, 305–307.

24. Bloch, K. (1969). Enzymatic synthesis of mono-unsaturated fatty acids. *Acc. Chem. Res.* **2**, 193–202.

25. Brock, D. J. H., Kass, L. R. & Bloch, K. (1967). $\beta$-Hydroxydecanoyl thioester dehydrase; II. Mode of action. *J. Biol. Chem.* **242**, 4432–4440.

26. Kass, L. R. & Bloch, K. (1967). The enzymatic synthesis of unsaturated fatty acids in *Escherichia coli. Proc. Natl. Acad. Sci. USA* **58**, 1168–1173.

27. Schwab, J. M., Ho, C.-K., Li, W.-B., Townsend, C. A. & Salituro, G. M. (1986). $\beta$-Hydroxydecanoyl thioester dehydrase. Complete characterization of the fate of the "suicide" substrate 3-decynoyl-NAC. *J. Am. Chem. Soc.* **108**, 5309–5316.

28. Walsh, C. (1982). Suicide substrates: Mechanism-based enzyme inactivators. *Tetrahedron* **38**, 871–909.

29. Rando, R. R. (1980). New modes of enzyme inactivator design. *Trends in Pharmacol. Sci.* **1**, 168–171.

30. Kalman, T. I. (1981). Enzyme inhibition as a source of new drugs. *Drug Dev. Res.* **1**, 311–328.

31. Brodbeck, U. (ed.) (1980). *Enzyme Inhibitors.* Verlag Chemie, Weinheim.

32. Bey, P. (1980). Design of enzyme-activated irreversible inhibitors of pyridoxal phosphate dependent enzymes. *Chemistry and Industry* 139–144.

33. Fowler, L. J. & John, R. A. (1972). Active site-directed irreversible inhibition of rat brain 4-aminobutyrate aminotransferase by ethanolamine-$O$-sulfate *in vitro* and *in vivo. Biochem. J.* **130**, 569–573.

34. Metcalf, B. W. (1978). Inhibitors of GABA metabolism. *Biochem. Pharmacol.* **28**, 1705–1712.

35. Jung, M. J., Lippert, B., Metcalf, B. W., Schlechter, P. J., Böhlen, P. & Sjoerdsma, A. (1977). The effect of 4-amino-hex-5-ynoic acid ($\gamma$-acetylenic GABA, $\gamma$-ethynyl GABA) a catalytic inhibitor of GABA transaminase on brain GABA metabolism *in vivo. J. Neurochem.* **28**, 717–723.

36. Metcalf, B. W., Lippert, B. & Casara, P. (1978). Enzyme-activated irreversible inhibition of transaminases. In *Enzyme-activated Irreversible Inhibitors* (Seiler, N., Jung, M. J. & Koch-Weser, J., eds), pp. 123–133. Elsevier/North Holland, Amsterdam.

37. Metcalf, B. W. & Sjoerdsma, A. (1981). Suicide inhibition of GABA-transaminase. In *Molecular Basis of Drug Action* (Singer, T. & Ondarza, P. N., eds), pp. 119–133. Elsevier/North Holland, Amsterdam.

38. Lippert, B., Metcalf, B. W. & Resvick, R. J. (1982). Enzyme-activated irreversible inhibition of rat and mouse brain 4-aminobutyric acid-$\alpha$-ketoglutarate transaminase by 5-fluoro-4-oxo-pentanoic acid. *Biochem. Biophys. Res. Comm.* **108**, 146–152.

39. Metcalf, B. & Sjoerdsma, A. (1979). The microscopic reversibility principle in enzyme inhibition. In *Drug Action and Design—Mechanism-based Enzyme Inhibitors* (Kaplan, T. I., ed.), pp. 61–73. Elsevier/North Holland, Amsterdam.

40. Lippert, B., Metcalf, B. W., Jung, M. J. & Casara, P. (1977). 4-Amino-hex-5-enoic acid, a selective catalytic inhibitor of 4-aminobutyric acid aminotransferase in mammalian brain. *Eur. J. Biochem.* **74**, 441–445.

41. Jung, M. J., Lippert, B., Metcalf, B. W., Böhlen, P. & Schechter, P. J. (1977). $\gamma$-vinyl GABA (4-amino-hex-5-enoic acid), a new selective irreversible inhibitor of GABA-T: Effects on brain GABA metabolism in mice. *J. Neurochem.* **29**, 797–802.

42. Silverman, R. B. & Invergo, B. J. (1986). Mechanism of inactivation of $\gamma$-aminobutyrate aminotransferase by 4-amino-5-fluoropentanoic acid. First example of an enamine mechanism for a $\gamma$-amino acid, with a partition ratio of 0. *Biochemistry* **25**, 6817–6820.

43. Silverman, R. B. & Levy, M. A. (1981). Mechanism of inactivation of $\gamma$-aminobutyric acid–$\alpha$-ketoglutaric acid aminotransferase by 4-amino-5-halopentanoic acids. *Biochemistry* **20**, 1197–1203.

44. Silverman, R. B., Durkee, S. C. & Invergo, B. J. (1986). 4-amino-2-(substituted methyl)-2-butenoic acids: Substrates and potent inhibitors of $\gamma$-aminobutyric acid aminotransferase. *J. Med. Chem.* **29**, 764–770.

45. Bey, P., Jung, M. J., Gerhart, F., Schirlin, D., Van Dorsselaer, V. & Casara, P. (1981). $\omega$-Fluoromethyl analogues of $\omega$-amino acids as irreversible inhibitors of 4-aminobutyrate; 2-oxoglutarate aminotransferase. *J. Neurochem.* **37**, 1341–1344.

46. Schechter, P. J. (1986). Vigabatrin. In *New Anticonvulsant Drugs* (Meldrum, B. S. & Porter, R. J., eds), pp. 265–275. John Libbey, London.

47. Boeker, E. A. & Snell, E. E. (1972). Amino acid decarboxylases. In *The Enzymes* (Boyer, P., ed.), Vol. VI, pp. 217–253. Academic Press, London.

48. Bey, P. (1978). Substrate-induced irreversible inhibition of $\alpha$-amino acid decarboxylases. Application to glutamate, aromatic-L-$\alpha$-amino acid and ornithine decarboxylases. In *Enzyme-activated Irreversible Inhibitors* (Seiler, N., Jung, M. J. & Koch-Weser, J., eds), pp. 27–42. Elsevier/North Holland, Amsterdam.

49. Metcalf, B. W., Bey, P., Danzin, C., Jung, M., Casara, P. & Vevert, J. P. (1978). Catalytic irreversible inhibition of mammalian ornithine decarboxylase

(EC 4.1.1.17) by substrate and product analogues. *J. Am. Chem. Soc.* **100**, 2551–2553.

50. Palfreyman, M. G., Danzin, C., Bey, P., Jung, M. J., Ribereau-Gayon, G., Aubry, M., Vevert, J. P. & Sjoerdsma, A. (1978). α-Difluoromethyl-dopa, a new enzyme-activated irreversible inhibitor of aromatic L-amino acid decarboxylase. *J. Neurochem.* **31**, 927–932.

51. Ribereau-Gayon, G., Palfreyman, M. G., Zreika, M., Wagner, J. & Jung, M. J. (1980). Irreversible inhibition of aromatic-L-amino acid decarboxylase by α-difluoromethyl-dopa and metabolism of the inhibitor. *Biochem. Pharmacol.* **29**, 2465–2469.

52. Kollonitsch, J., Patchett, A. A., Marburg, S., Maycock, A. L., Perkins, L. M., Doldouras, G. A., Duggan, D. E. & Aster, S. D. (1978). Selective inhibitors of biosynthesis of aminergic neurotransmitters. *Nature* **274**, 906–908.

53. Hornsberger, J. M. (1984). Ph.D. thesis, Université Louis Pasteur, Strasbourg, France.

54. Jung, M. J. (1986). Substrates and inhibitors of aromatic amino acid decarboxylase. *Bioorganic Chem.* **4**, 429–443.

55. Jung, M. J., Palfreyman, M. G., Wagner, J., Bey, P., Ribereau-Gayon, G., Zreika, M. & Koch-Weser, J. (1979). Inhibition of monoamine synthesis by irreversible blockade of aromatic aminoacid decarboxylase with α-monofluoromethyl-dopa. *Life Sci.* **24**, 1037–1042.

56. Bouclier, M., Jung, M. J. & Gerhart, F. (1983). α-Fluoromethyl histidine. Inhibition of histidine decarboxylase in pylorus ligated rat. *Biochem. Pharmacol.* **32**, 1553–1556.

57. Bouclier, M., Jung, M. J. & Gerhart, F. (1983). Effect of prolonged inhibition of histidine decarboxylase on tissue histamine concentrations. *Experientia* **39**, 1303–1305.

58. Kuo, D. & Rando, R. R. (1981). Irreversible inhibition of glutamate decarboxylase by α-(fluoromethyl)glutamic acid. *Biochemistry* **20**, 506–511.

59. Chrystal, E., Bey, P. & Rando, R. R. (1979). The irreversible inhibition of brain L-glutamate-1-decarboxylase by (2*RS*,3*E*)-2-methyl-3,4-didehydro-glutamic acid. *J. Neurochem.* **32**, 1501–1507.

60. Bey, P. (1985). Le fluor dans l'inhibition suicide de reactions enzymatiques. In *Le Fluor et les Materiaux Fluores* (Huong, P. V., ed.), pp. 695–702. Masson, Paris.

61. Walsh, C. (1983). Fluorinated substrate analogs. Route of metabolism and selective toxicity. In *Advances in Enzymology* (Meister, A., ed.), pp. 197–289. Wiley, New York.

62. Maycock, A. L., Aster, S. P. & Patchett, A. A. (1978). Studies with inhibitors of aromatic amino acid decarboxylase. In *Enzyme-activated Irreversible Inhibitors* (Seiler, N., Jung, M. J. & Koch-Weser, J., eds), pp. 211–220. Elsevier/North Holland, Amsterdam.

63. Ribereau-Gayon, G., Danzin, C., Palfreyman, M. G., Aubrey, M., Wagner, J., Metcalf, B. W. & Jung, M. J. (1979). *In vitro* and *in vivo* effects of α-acetylenic DOPA and α-vinyl DOPA on aromatic L-aminoacid decarboxylase. *Biochem. Pharmacol.* **28**, 1331–1335.

64. Jung, M. J., Metcalf, B. W., Lippert, B. & Casara, P. (1978). Mechanism of the stereospecific irreversible inhibition of bacterial glutamic acid decarboxylase by (*R*)-(−)-4-aminohex-5-ynoic acid, an analogue of 4-aminobutyric acid. *Biochemistry* **17**, 2628–2632.

65. Bouclier, M., Jung, M. J. & Lippert, B. (1979). Stereochemistry of reactions

catalyzed by mammalian brain L-glutamate-1-carboxylase and 4-amino-butyrate:2-oxoglutarate amino transferase. *Eur. J. Biochem.* **98**, 363–368.

66. Danzin, C., Claverie, N. & Jung, M. J. (1984). Stereochemistry of the inactivation of 4-aminobutyrate:2-oxoglutarate amino transferase and L-glutamate-1-carboxylase by 4-aminohex-5-ynoic acid enantiomers. *Biochem. Pharmacol.* **33**, 1741–1746.

67. Danzin, C., Bey, P., Schirlin, D. & Claverie, N. (1982). $\alpha$-Monofluoromethyl and $\alpha$-difluoromethyl putrescine as ornithine decarboxylase inhibitors: *In vitro* and *in vivo* biochemical properties. *Biochem. Pharmacol.* **31**, 3871–3878.

68. Bey, P., Danzin, C. & Jung, M. (1987). Inhibition of basic amino acid decarboxylases involved in polyamine biosynthesis. In *Inhibition of Polyamine Biosynthesis: Biological Significance and Basis for New Therapies* (McCann, P. P., Pegg, A. E. & Sjoerdsma, A., eds). pp. 1–31. Academic Press, London.

69. Floss, H. G. & Vederas, J. C. (1982). In *Stereochemistry* (Tamm, C., ed.), pp. 161–199. Biomedical Press, New York.

70. Casara, P., Danzin, C., Metcalf, B. W. & Jung, M. J. (1982). Stereospecific irreversible inhibition of mammalian (*S*)-ornithine decarboxylase by (*R*)-(−)-hex-5-yne-1,4-diamine. *J. Chem. Soc., Chem. Commun.*, 1190–1192.

71. Bey, P., Metcalf, B., Jung, M. J., Fozard, J. & Koch-Weser, J. (1982). Substrate-induced irreversible inhibition of enzymes in drug research. In *Strategy in Drug Research* (Keverling Buisman, J. A., ed.), pp. 89–106. Elsevier/North Holland, Amsterdam.

72. Bey, P., Gerhart, F. & Jung, M. (1986). Synthesis of (*E*)-4-amino-2,5-hexadienoic acid and (*E*)-4-amino-5-fluoro-2-pentenoic acid. Irreversible inhibitors of 4-aminobutyrate-2-oxoglutarate aminotransferase. *J. Org. Chem.* **51**, 2835–2838.

73. Bey, P., Gerhart, F., Van Dorsselaer, V. & Danzin, C. (1983). $\alpha$-(Fluoromethyl)-dehydroornithine and $\alpha$-(fluoromethyl)-dehydroputrescine analogues as irreversible inhibitors of ornithine decarboxylase. *J. Med. Chem.* **26**, 1551–1556.

74. Relyea, N. & Rando, R. R. (1975). Potent inhibition of ornithine decarboxylase by $\beta,\gamma$-unsaturated substrate analogs. *Biochem. Biophys. Res. Commun.* **67**, 392–402.

75. Silverman, R. B. & Levy, M. A. (1981). Substituted 4-aminobutanoic acids. Substrates for $\gamma$-aminobutyric acid $\alpha$-ketoglutaric acid aminotransferase. *J. Biol. Chem.* **256**, 11 565–11 568.

76. McDonald, I. A., Lacoste, J. M., Bey, P., Wagner, J., Zreika, M. & Palfreyman, M. G. (1986). Dual enzyme-activated irreversible inhibition of monoamine oxidase. *Bioorganic Chem.* **14**, 103–118.

77. McDonald, I. A., Palfreyman, M. G., Jung, M. J. & Bey, P. (1985). Synthesis of (*E*)-$\beta$-fluoromethylene glutamic acid. *Tetrahedron Lett.* **26**, 4091–4092.

78. Likos, J. J., Ueno, H., Feldhaus, R. W. & Metzler, D. E. (1982). A novel reaction of the coenzyme of glutamate decarboxylase with L-serine *O*-sulfate. *Biochemistry* **21**, 4377–4386.

79. Badet, B., Roise, D. & Walsh, C. T. (1984). Inactivation of the *dadB Salmonella typhimurium* alanine racemase by D and L isomers of $\beta$-substituted alanines: Kinetics, stoichiometry active site peptide sequencing, and reaction mechanism. *Biochemistry* **23**, 5188–5194.

80. Hayashi, H., Tanase, S. & Snell, E. E. (1986). Pyridoxol 5′-phosphate-dependent histidine decarboxylase. Inactivation by $\alpha$-fluoromethyl histidine and

comparative sequences of the inhibitor and coenzyme binding sites. *J. Biol. Chem.* **261**, 11 003–11 009.

81. Murphy, D. L., Garrick, N. A., Anlakh, C. S. & Cohen, R. M. (1984).1 New contributions from basic science to understanding the effects of monoamine oxidase inhibiting antidepressants. *J. Clin. Psychiatry* **45**, 37–43.

82. McDonald, I. A., Bey, P. & Palfreyman, M. G. (1987). Monoamine oxidase inhibitors. In *Design of Enzyme Inhibitors as Drugs* (Sandler, M. & Smith, J., eds). Oxford University Press, Oxford (in press).

83. Salach, J. I., Detmer, K. & Youdim, M. B. H. (1979). The reaction of bovine and rat liver monoamine oxidase with [$^{14}$C]-clorgyline and [$^{14}$C]-deprenyl. *Mol. Pharmacol.* **16**, 234–241.

84. Krantz, A., Kokel, B., Sachdeva, Y. P., Salach, J., Detmer, K., Claesson, A. & Sahlberg, C. (1979). Inactivation of mitochondrial monoamine oxidase by $\beta,\gamma,\delta$-allenic amines. In *Monoamine Oxidase: Structure, Function, and Altered Functions* (Singer, T. P., Von Korff, R. W. & Murphy, D. L., eds), pp. 51–70. Academic Press, London.

85. Bey, P., Fozard, J., Lacoste, J. M., McDonald, I. A., Zreika, M. & Palfreyman, M. G. (1984). ($E$)-2-(3,4-dimethoxyphenyl)-3-fluoroallylamine: A selective, enzyme-activated inhibitor of type B monoamine oxidase. *J. Med. Chem.* **27**, 9–10.

86. Zreika, M., McDonald, I. A., Bey, P. & Palfreyman, M. G. (1984). MDL 72,145, an enzyme-activated irreversible inhibitor with selectivity for monoamine oxidase type B. *J. Neurochem.* **43**, 448–454.

87. Paech, C., Salach, J. I. & Singer, T. P. (1979). Suicide inactivation of monoamine oxidase by *trans*-phenylcyclopropylamine. In *Monoamine Oxidase: Structure, Function, and Altered Functions* (Singer, T. P., Von Korff, R. W. & Murphy, D. L., eds), pp. 39–50. Academic Press, London.

88. McDonald, I. A., Lacoste, J. M., Bey, P., Palfreyman, M. G. & Zreika, M. (1985). Enzyme-activated irreversible inhibitors of monoamine oxidase: Phenylallylamine structure–activity relationship. *J. Med. Chem.* **28**, 186–193.

89. McDonald, I. A., Palfreyman, M. G., Zreika, M. & Bey, P. (1986). ($Z$)-2-(2,4-dichlorophenoxy)methyl-3-fluoroallylamine (MDL 72,638): A clorgyline analogue with surprising selectivity for monoamine oxidase type B. *Biochem. Pharmacol.* **35**, 349–351.

90. Palfreyman, M. G., McDonald, I. A., Bey, P., Schechter, P. J. & Sjoerdsma, A. (1986). Design and early clinical evaluation of selective inhibitors of monoamine oxidase. ACNP Proceedings, Abstract 30, 8–12 December, 1986.

91. McDonald, I. A., Lacoste, J. M., Bey, P., Wagner, J., Zreika, M. & Palfreyman, M. G. (1984). ($E$)-$\beta$-(Fluoromethylene)-*m*-tyrosine: A substrate for aromatic L-amino acid decarboxylase liberating an enzyme-activated irreversible inhibitor of monoamine oxidase. *J. Am. Chem. Soc.* **106**, 3354–3356.

92. Palfreyman, M. G., McDonald, I. A., Fozard, J. R., Mely, Y., Sleight, A. J., Zreika, M., Wagner, J., Bey, P. & Lewis, P. J. (1985). Inhibition of monoamine oxidase selectively in brain monoamine nerves using the bio-precursor ($E$)-$\beta$-fluoromethylene-*m*-tyrosine (MDL 72,394): A substrate for aromatic L-amino acid decarboxylase. *J. Neurochem.* **45**, 1850–1860.

93. Jung, M. J., Hornsberger, J.-M., McDonald, I. A., Fozard, J. R. & Palfreyman, M. G. (1985). Bio-precursor approach to site-selective enzyme inhibition. In *Drug Targeting* (Buri, P. & Gumma, A., eds). pp. 165–178. Elsevier/North Holland, Amsterdam.

94. Palfreyman, M. G., McDonald, I., Zreika, M. & Fozard, J. R. (1984). MDL 72,394: The prodrug approach to brain selective MAO inhibition. In *Monoamine Oxidase and Disease* (Tipton, K. F., Dostert, P. & Strolin Benedetti, M., eds), pp. 561–562. Academic Press, London.
95. Mely, Y., Palfreyman, M. G. & Zreika, M. (1984). *In vitro* evidence for the neuronal selectivity of the MAO inhibiting prodrug, MDL 72,394. *J. Pharm. Pharmacol.* **36** (Workshop Suppl. 39W).
96. Jung, M. J., Hornsberger, J. M., Gerhart, F. & Wagner, J. (1984). Inhibition of aromatic amino acid decarboxylase and depletion of biogenic amines in brain of rats treated with alpha-mono-fluoromethyl-*p*-tyrosine: Similitudes and differences with the effects of alpha monofluoromethyl-dopa. *Biochem. Pharmacol.* **33**, 327–330.
97. Bey, P., Bolkenius, F. N., Seiler, N. & Casara, P. (1985). *N*-2,3-Butadienyl-1,4-butanediamine derivatives: Potent irreversible inactivators of mammalian polyamine oxidase. *J. Med. Chem.* **28**, 1–2.
98. Bolkenius, F. N., Bey, P. & Seiler, N. (1985). Specific inhibition of polyamine oxidase *in vivo* as a method for the elucidation of its physiological role. *Biochim. Biophys. Acta* **838**, 69–76.
99. Lobenstein-Verbeck, C. L., Jongejon, J. A., Frank, J. & Duine, J. A. (1984). Bovine serum amine oxidase: a mammalian enzyme having covalently bound PQQ as prosthetic group. *FEBS Lett.* **170**, 305–309.
100. Williamson, P. R., Moog, R. S., Dooley, D. M. & Kagan, H. M. (1986). Evidence for pyrroloquinoline quinone as the carbonyl cofactor in lysyl oxidase by absorption and resonance ramon spectroscopy. *J. Biol. Chem.* **261**, 16302–16305.
101. Hevey, R. C., Babson, J., Maycock, A. L. & Abeles, R. H. (1973). Highly specific enzyme inhibitors. Inhibition of plasma amine oxidase. *J. Am. Chem. Soc.* **95**, 6125–6126.
102. Maycock, A., Suva, R. & Abeles, R. (1975). Novel inactivators of plasma amine oxidase. *J. Am. Chem. Soc.* **97**, 5613–5614.
103. Mangold, J. B. & Klinman, J. P. (1984). Mechanism-based inactivation of dopamine-β-monooxygenase by β-chlorophenethylamine. *J. Biol. Chem.* **259**, 7772–7779.
104. Columbo, G., Rajashekhar, B., Giedroc, D. P. & Villafranca, J. J. (1984). Alternate substrates of dopamine β-hydroxylase. *J. Biol. Chem.* **259**, 1593–1600.
105. Rajashekhar, B., Fitzpatrick, P. F., Columbo, G. & Villafranca, J. J. (1984). Synthesis of several 2-substituted 3-(*p*-hydroxyphenyl)-1-propenes and their characterization as mechanism-based inhibitors of dopamine β-hydroxylase. *J. Biol. Chem.* **259**, 6925–6930.
106. Columbo, G. & Villafranca, J. J. (1984). An acetylenic mechanism-based inhibitor of dopamine β-hydroxylase. *J. Biol. Chem.* **259**, 15017–15020.
107. May, S. W., Mueller, P. W., Padgette, S. R., Herman, H. H. & Philips, R. S. (1983). Dopamine-β-hydroxylase: Suicide inhibition by the novel olefinic substrate, 1-phenyl-1-aminomethylene. *Biochem. Biophys. Res. Commun.* **110**, 161–168.
108. Creveling, C. R., Vander Schoot, J. B. & Udenfriend, S. (1962). Phenethylamine isosteres as inhibitors of dopamine β-oxidase. *Biochem. Biophys. Res. Commun.* **8**, 215–226.
109. Bargar, T. M., Broersma, R. J., Creemer, L. C., McCarthy, J. R., Hornsberger, J.-M., Palfreyman, M. G., Wagner, J. & Jung, M. J. (1986).

Unsaturated heterocyclic amines as potent time-dependent inhibitors of dopamine β-hydroxylase. *J. Med. Chem.* **29**, 315–317.

110. Padgette, S. R., Wimalasena, K., Herman, H. H., Sirimanne, S. R. & May, S. W. (1985). Olefin oxygenation and *N*-dealkylation by dopamine β-monooxygenase: Catalysis and mechanism-based inhibition. *Biochemistry* **24**, 5826–5839.

111. Hardy, L. W., Finer-Moore, J. S., Montfort, W. R., Jones, M. O., Santi, D. V. & Stroud, R. M. (1987). Atomic structure of thymidylate synthase: target for rational drug design. *Science* **235**, 448–455.

112. Parks, D. A. & Granger, D. N. (1986). Xanthine oxidase: Biochemistry, distribution and biology. *Acta Physiol. Scand.* **548** (Suppl.), 87–99.

113. Macrum, B. L., Nelson, E. E. & Povzhitkov, M. M. (1986). Beneficial effects of allopurinol on survival and infarct size following coronary artery ligation in the rat. *Br. J. Pharmacol.* **89**, 544 pp.

114. Cha, S., Agarwell, R. P. & Parks, R. E. (1975). Tight-binding inhibitors, II: Non-steady state nature of inhibition of milk xanthine oxidase by allopurinol and alloxanthine and of human erythrocytic adenosine deaminase by coformycin. *Biochem. Pharmacol.* **24**, 2187–2197.

115. Rando, R. R. (1977). Mechanism of the irreversible inhibition of γ-aminobutyric acid-α-ketoglutaric acid transaminase by the neurotoxin gabaculine. *Biochemistry* **16**, 4604–4610.

116. Metcalf, B. W. & Jung, M. J. (1979). Molecular basis for the irreversible inhibition of 4-aminobutyric acid:2-oxoglutarate and L-ornithine:2-oxoacid aminotransferases by 3-amino-1,5-cyclohexadienyl carboxylic acid (Isogabaculine). *Mol. Pharmacol.* **16**, 539–545.

117. Lippert, B., Resvick, R., Burkhart, J., Holbert, G., Adams, J. & Metcalf, B. (1985). Aromatization: A driving force for inhibition of pyridoxal phosphate-dependent enzymes. *Fed. Proc.* **44**, 1399.

118. Ortiz de Montellano, P. R. & Correia, M. A. (1983). Suicidal destruction of cytochrome P-450 during oxidative drug metabolism. *Ann. Rev. Pharmacol. Toxicol.* **23**, 481–503.

119. Metcalf, B. W., Wright, C. L., Burkhart, J. P. & Johnston, J. O. (1981). Substrate-induced inactivation of aromatase by allenic and acetylenic steroids. *J. Am. Chem. Soc.* **103**, 3221–3222.

120. Blohm, T. R., Metcalf, B. W., Laughlin, M. E., Sjoerdsma, A. & Schatzman, G. L. (1980). Inhibition of testosterone 5α-reductase by a proposed enzyme-activated, active site-directed inhibitor. *Biochem. Biophys. Res. Commun.* **95**, 273–280.

121. Walker, B. (1986). Mechanism-based proteinase inhibitors: A critical review. *Biochem. Soc. Trans.* **14**, 397–399.

122. Turner, A. J . (1986). Strategies for the inhibition of neuropeptide-metabolizing enzymes. *Biochem. Soc. Trans.* **14**, 399–401.

123. Fuller, R. W., Slater, I. M. & Mills, J. (1979). The development of *n*-cyclopropyl-arylalkylamines as monoamine oxidase inhibitors. In *Monoamine Oxidase: Structure, Function, and Altered Functions* (Singer, T. P., Von Korff, R. W. & Murphy, D. L., eds), pp. 317–331. Academic Press, London.

124. Gardner, C. R. & Richard, M. H. (1981). Use of a D,L-α-monofluoromethyl-dopa to distinguish subcellular pools of aromatic amino acid decarboxylase in mouse brain. *Brain Res.* **216**, 291–298.

125. Pegg, A. E. (1986). Recent advances in the biochemistry of polyamines in eukaryotes. *Biochem. J.* **234**, 249–262.

126. Parsons, B. & Rainbow, T. C. (1984). High affinity binding sites for [$^3$H]MPTP may correspond to monoamine oxidase. *Eur. J. Pharmacol.* **102**, 375–377.
127. Gomez, N., Unzeta, M., Tipton, K. F., Anderson, M. C. & Carroll, A.-M. (1986). Determination of monoamine oxidase concentrations in rat liver by inhibitor binding. *Biochem. Pharmacol.* **35**, 4467–4472.
128. Callingham, B. A. & Parkinson, D. (1979). Tritiated pargyline binding to rat liver mitochondrial MAO. In *Monoamine Oxidase: Structure, Function, and Altered Functions* (Singer, T. P., Von Korff, R. W. & Murphy, D. L., eds), pp. 81–86. Academic Press, London.
129. Cubells, J. F., Blanchard, J. S., Smith, D. M. & Makman, M. H. (1986). *In vivo* action of enzyme-activated irreversible inhibitors of glutamic acid decarboxylase and γ-aminobutyric acid transaminase in retina vs. brain. *J. Pharmacol. Exp. Ther.* **238**, 508–516.
130. Fowler, J. S., MacGregor, R. R., Wolf, A. P., Arnett, C. D., Dewey, S. L., Schlyer, D., Christman, D., Logan, J., Smith, M., Sachs, H., Aquilonius, S. M., Bjurling, P., Halldin, C., Hastvig, P., Leenders, K. L., Lundqvist, M., Oreland, L., Stalnacke, C. G. & Longstrom, B. (1987). Mapping human brain monoamine oxidase A and B with $^{11}$C-labeled suicide inactivator and PET. *Science* **235**, 481–485.
131. Salach, J. I. & Detmer, K. (1979). Chemical characterization of monoamine oxidase from human placental mitochondria. In *Monoamine Oxidase: Structure, Function, and Altered Functions* (Singer, T. P., Von Korff, R. W. & Murphy, D. L., eds), pp. 121–128. Academic Press, London.
132. Pintar, J. E., Cawthon, R. M., Castro Costa, M. R. & Breakefield, X. O. (1979). A search for structural differences in MAO: Electrophoretic analysis of $^3$H-pargyline labeled proteins. In *Monoamine Oxidase: Structure, Function, and Altered Functions* (Singer, T. P., Von Korff, R. W. & Murphy, D. L., eds), pp. 185–196. Academic Press, London.
133. Cawthon, R. M., Pintar, J. E., Haseltine, F. P. & Breakefield, X. O. (1981). Differences in the structure of A and B forms of human monoamine oxidase. *J. Neurochem.* **37**, 363–372.
134. Smith, D., Filipowicz, C. & McCauley, R. (1985). Monoamine oxidase A and monoamine oxidase B activities are catalyzed by different proteins. *Biochim. Biophys. Acta* **831**, 1–7.
135. Churchich, J. E. & Moses, U. (1981). 4-Amino-butyrate aminotransferase: the presence of nonequivalent binding sites. *J. Biol. Chem.* **256**, 1101–1104.
136. Burnett, G., Marcotte, P. & Walsh, C. (1980). Mechanism-based inactivation of pig heart L-alanine transaminase by L-propargylglycine. *J. Biol. Chem.* **255**, 3487–3491.
137. Jancsar, S. M. & Leonard, B. E. (1984). Changes in neurotransmitter metabolism following olfactory bulbectomy in the rat. *Prog. Neuropsychopharmacol. and Biol. Psychiat.* **8**, 263–269.
138. Gale, K. & Iadarola, M. J. (1980). Seizure protection and increased nerve-terminal GABA: Delayed effects of GABA transaminase inhibition. *Science* **208**, 288–291.
139. Forchetti, C. M., Marco, E. J. & Meek, J. L. (1982). Serotonin and γ-aminobutyric turnover after injection into the median raphe of substance P and D-ala-met-enkephalin-amide. *J. Neurochem.* **38**, 1336–1341.
140. Tozer, T. N., Neff, N. M. & Brodie, B. B. (1966). Application of steady-state

kinetics to the synthesis rate and turnover time of serotonin in the brain of normal and reserpine-treated rats. *J. Pharmacol. Exp. Ther.* **153**, 177–182.

141. Palfreyman, M. G., Zreika, M., Arbogast, R. & Wagner, J. (1984). A method for measuring monoamine turnover in animals using an irreversible inhibitor of aromatic L-amino acid decarboxylase, DL-$\alpha$-monofluoromethyl-dopa. *J. Pharmacol. Methods* **11**, 239–251.

142. Mamont, P. S., Duchesne, M.-C., Grove, J. & Bey, P. (1978). Antiproliferative properties of DL-$\alpha$-difluoromethylornithine in cultured cells. A consequence of the irreversible inhibition of ornithine decarboxylase. *Biochem. Biophys. Res. Commun.* **81**, 58–66.

143. Prakash, N. J., Schechter, P. J., Mamont, P. S., Grove, J., Koch-Weser, J. & Sjoerdsma, A. (1980). Inhibition of EMT6 tumor growth by interference with polyamine biosynthesis; effects of $\alpha$-difluoromethylornithine, an irreversible inhibitor of ornithine decarboxylase. *Life Sci.* **26**, 181–194.

144. Bacchi, C. J., Nathan, H. C., Hutner, S. H., McCann, P. P. & Sjoerdsma, A. (1980). Polyamine metabolism: A potential therapeutic target in trypanosomes. *Science* **210**, 332–334.

145. Sjoerdsma, A. & Schlechter, P. J. (1984). Chemotherapeutic implications of polyamine biosynthesis inhibition. *Clin. Pharmacol. Ther.* **35**, 287–300.

146. Sjoerdsma, A., Golden, J. A., Schechter, P. J., Barlow, J. L. R. & Santi, D. V. (1984). Successful treatment of lethal protozoal infections with the ornithine decarboxylase inhibitor, $\alpha$-difluoromethylornithine. *Trans. Assoc. Am. Phys.* **97**, 70–79.

147. Schechter, P. J., Barlow, J. L. R. & Sjoerdsma, A. (1987). Clinical aspects of inhibition of ornithine decarboxylase with emphasis on therapeutic trials of eflorinithine (DFMO) in cancer and protozoan diseases. In *Inhibition of Polyamine Metabolism: Biological Significance and Basis for New Therapies* (McCann, P. P., Pegg, A. E. & Sjoerdsma, A., eds), pp. 345–364. Academic Press, London.

148. McCann, P. P., Pegg, A. E. & Sjoerdsma, A. (eds) (1987). *Inhibition of Polyamine Metabolism: Biological Significance and Basis for New Therapies.* Academic Press, London.

149. Sjoerdsma, A. (1981). Suicide enzyme inhibitors as potential drugs. *Clin. Pharmacol. Ther.* **30**, 3–22.

# Polyamines

## DAVID M. L. MORGAN

*Section of Vascular Biology, Medical Research Council Clinical Research Centre, Harrow, Middlesex HA1 3UJ, UK*

## I. Introduction

The polyamines spermine, spermidine and putrescine, although receiving scant attention in biochemical textbooks as yet, have a respectably long history. The formation of crystals of spermine phosphate in samples of human semen was first noted by Antony van Leeuwenhoek in 1678,[1] and subsequently rediscovered several times.[2] The structure of spermine was finally settled by its synthesis in 1926,[3] and spermidine was synthesized the following year.[4] Putrescine was isolated in 1879[5] and synthesized seven years later.[6] For accounts of the early work on polyamines the readers is referred to the monographs by Cohen[7] and Bachrach,[2] and several reviews.[8–10]

Why are the polyamines of interest? Spermidine and putrescine are ubiquitous in living organisms and spermine is present in virtually all cells of higher eukaryotes, often in millimolar amounts. Their concentrations vary with transit through the cell cycle (Fig. 1) and induction of ornithine decarboxylase, a key enzyme in polyamine biosynthesis, is a very early event

ESSAYS IN BIOCHEMISTRY Vol. 23
ISBN 0 12 158123-3

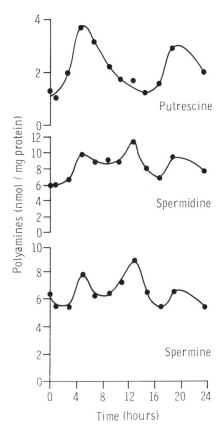

**Fig. 1.** Changes in polyamine content of HTC cells with passage through the cell cycle. Redrawn from Mamont.[189]

in cell proliferation. Intracellular polyamine levels usually increase before DNA, RNA or protein. Depletion of intracellular polyamine levels, by the use of inhibitors of their biosynthesis, results in a cessation of growth, which can be restored by the addition of exogenous polyamines to the inhibitor-treated cells.[11] The widespread occurrence of these compounds and the complexities of their regulation and function have engaged the interest of biochemists, organic chemists, microbiologists, virologists, plant physiologists, oncologists, and other clinicians. The increasing awareness of the importance of these compounds in cell growth and proliferation is shown by the rise in the number of papers reporting polyamine research listed in *Index Medicus* over the last 20 years, from 47 in 1966 to 386 in 1986. Consequently, the literature on polyamines is now large and diffuse, and therefore many

$$NH_2(CH_2)_4NH_2$$
putrescine

$$NH_2(CH_2)_3NH(CH_2)_4NH_2$$
spermidine

$$NH_2(CH_2)_3NH(CH_2)_4NH(CH_2)_3NH_2$$
spermine

**Fig. 2.** The three common polyamines.

references in this essay will be to reviews which can serve as a point of entry to the wider literature on a particular topic. No attempt has been made to make this essay comprehensive, rather it is intended to deal mainly with mammalian systems and to point out where other systems differ, to draw attention to areas of interest and to indicate how much we still do not know.

As can be seen in Fig. 2, the polyamines are low molecular weight aliphatic nitrogenous bases, and some of their properties are listed in Table 1. The differing dissociation constants of the amino groups in spermine and spermidine are noteworthy, as inductive effects should be negligible where the separation is greater than two methylene groups.[12] Some of the less common polyamines, found mainly in plants and microorganisms, are shown in Fig. 3.

| Diamines | $NH_2(CH_2)_3NH_2$ |
| | 1,3-diaminopropane |
| | $NH_2(CH_2)_5NH_2$ |
| | cadaverine |
| Triamines | $NH_2(CH_2)_3NH(CH_2)_3NH_2$ |
| | norspermidine |
| | (caldine) |
| | $NH_2(CH_2)_3NH(CH_2)_5NH_2$ |
| | aminopropylcadaverine |
| | $NH_2(CH_2)_4NH(CH_2)_4NH_2$ |
| | homospermidine |
| Tetra-amines | $NH_2(CH_2)_3NH(CH_2)_3NH(CH_2)_3NH_2$ |
| | norspermine |
| | (thermine) |
| | $NH_2(CH_2)_3NH(CH_2)_3NH(CH_2)_4NH_2$ |
| | thermospermine |
| | $NH_2(CH_2)_4NH(CH_2)_3NH(CH_2)_4NH_2$ |
| | canavalmine |
| Penta-amines | $NH_2(CH_2)_3NH(CH_2)_3NH(CH_2)_3NH(CH_2)_3NH_2$ |
| | caldopentamine |
| | $NH_2(CH_2)_3NH(CH_2)_3NH(CH_2)_3NH(CH_2)_4NH_2$ |
| | homocaldopentamine |

**Fig. 3.** Some of the less common polyamines, found mainly in plants and bacteria.[23,190]

## TABLE 1

### Some properties of the polyamines spermine, spermidine and putrescine

| Name | Synonyms | Chemical Abstracts Registry No. | Formula | $M_r$ | M Pt (hydro-chloride) | $pK^a$ |
|---|---|---|---|---|---|---|
| Putrescine | 1,4-diamino-butane | 110-60-1 | $C_4H_{12}N_2$ | 88·15 | >275 | 8·71 |
| Spermidine | $N$-(3-aminopropyl)-1,4-diaminobutane 4-azaoctane-1,8-diamine | 124-20-9 | $C_7H_{19}N_3$ | 145·25 | 256–258 | 10·96 9·91 8·51 |
| Spermine | $N,N'$-bis(3-amino-propyl)-1,4-diaminobutane | 71-44-3 | $C_{10}H_{26}N_4$ | 202·34 | 312–314 | 10·86 10·05 8·82 |

[a]Determined by potentiometric titration (spermidine,[192] spermine[193]).

## II. Biosynthesis and Interconversion of the Polyamines

A generalized pathway of polyamine biosynthesis is shown in Fig. 4. In mammalian and many other types of cell the initial, and at this stage rate-limiting, step is the decarboxylation of ornithine to form putrescine, catalysed by ornithine decarboxylase. The starting material (ornithine) is available from the plasma[13] and can also be formed within the cell by the action of arginase. It has been suggested[14] that arginase, which is much more widely distributed in tissues than other enzymes of the urea cycle, is present to ensure the availability of ornithine for polyamine production and that the decarboxylation of arginine can, therefore, be thought of as the first step in polyamine biosynthesis. Many microorganisms[15] and higher plants[16,17] can also synthesize putrescine from agmatine, formed by decarboxylation of arginine by arginine decarboxylase, an enzyme not present in mammalian cells or in many lower eukaryótes but widespread in higher plants. Agmatine is then hydrolysed by agmatinase to form putrescine, with the elimination of urea (Fig. 5). Some organisms possess both of these pathways, é.g. *Escherichia coli*. In plants additional routes exist from agmatine to putrescine; one via $N$-carbamoylputrescine is catalysed by $N$-carbamoylputrescine amidohydrolase (EC 3.5.1.?; the Enzyme Commission Handbook[18] is not clear on this point), and a multifunctional enzyme has been demonstrated in the grass pea *Lathyrus sativus*,[19] which catalyses the overall reaction of Fig. 6.

In a step common to most organisms spermidine is formed from putrescine by the addition of an aminopropyl group donated by decarboxylated $S$-adenosylmethionine, a reaction catalysed by the aminopropyltransferase spermidine synthase. Addition of a second aminopropyl moiety to spermidine, catalysed by a different aminopropyltransferase, spermine synthase, forms spermine.[20] The source of the aminopropyl groups is $S$-adenosylmethionine which is first decarboxylated by $S$-adenosylmethionine decarboxylase. The synthesis of spermidine and spermine is dependent on the availability of the aminopropyl donor, hence $S$-adenosylmethionine decarboxylase is also rate-limiting in polyamine biosynthesis. The methionine and adenosine moieties of the resulting methylthioadenosine (Fig. 7; a compound of interest in its own right[21,22]) are salvaged in mammalian cells by a series of reactions that is not yet fully understood.[11] In a number of bacteria L-aspartic-$\beta$-semialdehyde,[23] and not decarboxylated $S$-adenosyl-methionine, is the aminopropyl group donor in spermidine biosynthesis (Fig. 8), and this alternative pathway has been demonstrated also in *L. sativus*.[24]

In mammalian cells spermine and spermidine can be converted back to putrescine as shown in Fig. 4. The first step is the acetylation of an aminopropyl group of spermine, catalysed by the enzyme spermidine/sper-

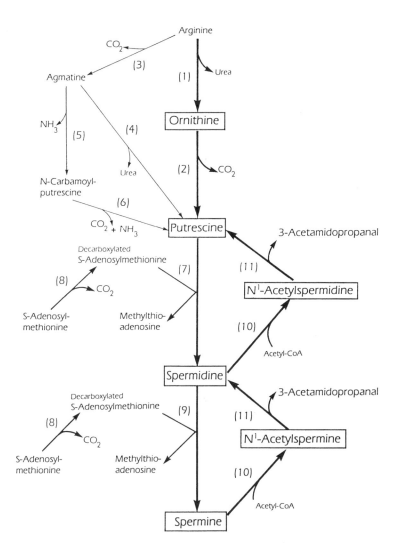

**Fig. 4.** A generalized pathway of polyamine biosynthesis and interconversion. Heavy arrows indicate the pathway in mammalian cells. The enzymes involved are: (1) arginase (EC 3.5.3.1); (2) ornithine decarboxylase (EC 4.1.1.17); (3) arginine decarboxylase (EC 4.1.1.19); (4) agmatinase (EC 3.5.3.11); (5) agmatine deiminase (EC 3.5.3.12); (6) $N$-carbamoylputrescine amidohydrolase (EC 3.5.1.?); (7) spermidine synthase (EC 2.5.1.16); (8) $S$-adenosylmethionine decarboxylase (EC 4.1.1.50); (9) spermine synthase (EC 2.5.1.22); (10) spermidine/spermine $N^1$-acetyltransferase; (11) polyamine oxidase (FAD-dependent). The last two enzymes have not yet had EC numbers assigned to them.

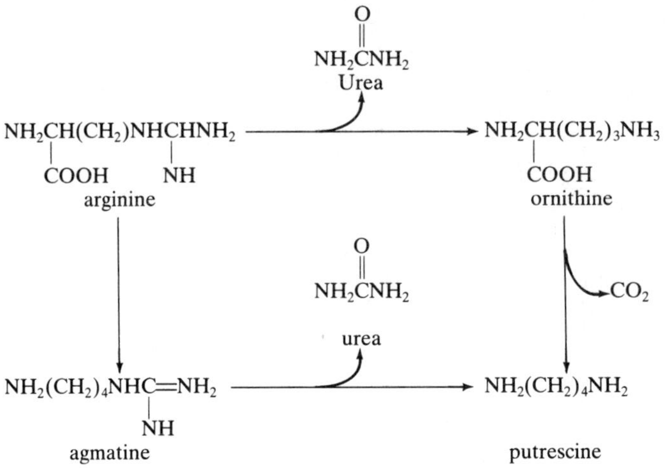

**Fig. 5.** Pathways for the biosynthesis of putrescine from arginine in bacteria and plants.

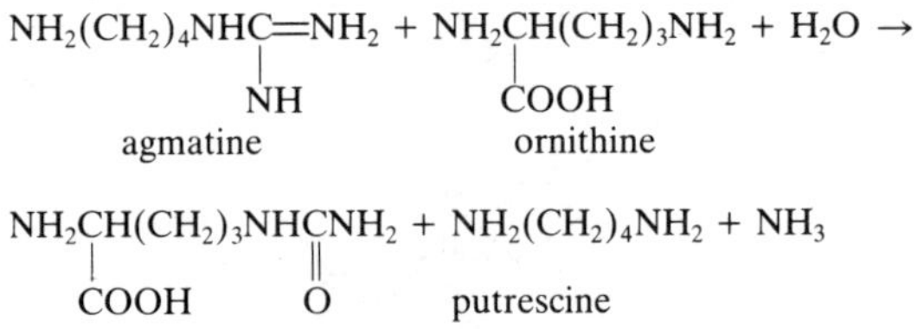

$$NH_2(CH_2)_4NHC{=}NH_2 + NH_2CH(CH_2)_3NH_2 + H_2O \rightarrow$$

$$\underset{\text{agmatine}}{NH} \qquad \underset{\text{ornithine}}{COOH}$$

$$NH_2CH(CH_2)_3NHCNH_2 + NH_2(CH_2)_4NH_2 + NH_3$$

$$\underset{COOH}{} \qquad \underset{O}{} \quad \text{putrescine}$$

**Fig. 6.** Synthesis of putrescine from ornithine and agmatine in *Lathyrus sativus*.

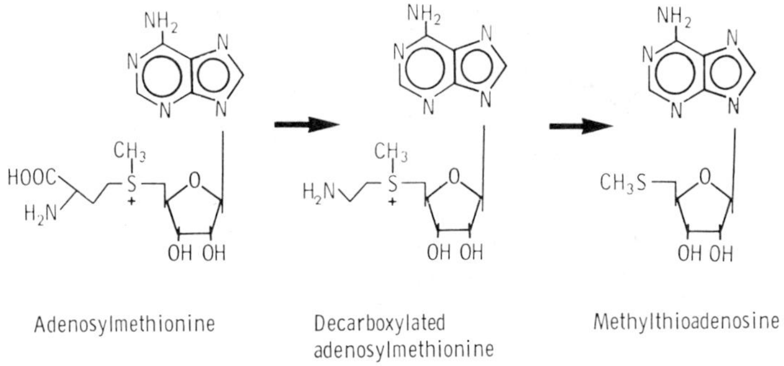

Adenosylmethionine        Decarboxylated              Methylthioadenosine
                          adenosylmethionine

**Fig. 7.** Pathway from *S*-adenosylmethionine to methylthioadenosine.

$$NH_2CHCH_2CHO + NH_2(CH_2)_4NH_2 \rightarrow NH_2(CH_2)_4NH(CH_2)_2CHNH_2 \rightarrow$$
$$\quad\;|$$
$$COOH \qquad\qquad\qquad\qquad\qquad\qquad\qquad\qquad COOH$$

aspartic            putrescine              "carboxyspermidine"
semialdehyde

$$NH_2(CH_2)_3NH(CH_2)_4NH_2 + CO_2$$
spermidine

**Fig. 8.** Synthesis of spermidine from aspartic semialdehyde in bacteria and plants.

$$NH_2(CH_2)_3NH(CH_2)_4NH(CH_2)_3NH_2 \xrightarrow{\text{Acetyl-CoA}}$$
spermine

$$CH_3CONH(CH_2)_3NH(CH_2)_4NH(CH_2)_3NH_2 \xrightarrow{\text{PAO}}$$
$N^1$-acetylspermine

$$NH_2(CH_2)_4NH(CH_2)_3NH_2 + CH_3CONH(CH_2)_2CHO$$
spermidine              3-acetamidopropanal

**Fig. 9.** Conversion of spermine to spermidine by acetylation by spermidine/spermine $N^1$-acetyl-transferase, then oxidation by polyamine oxidase (PAO).

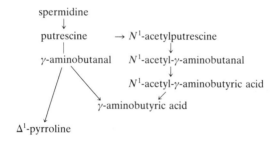

**Fig. 10.** Possible pathways of putrescine metabolism.[25,168]

$$CH_3CONH(CH_2)_2CHO \rightarrow CH_3CONH(CH_2)_2COOH \rightarrow$$
3-acetamidopropanal        3-acetamidopropanoic
acid

$$NH_2(CH_2)_2COOH$$
3-aminopropanoic
acid
($\beta$-alanine)

**Fig. 11.** Suggested route for the formation of $\beta$-alanine from 3-acetamidopropanal.[25]

mine $N^1$-acetyltransferase, to give $N^1$-acetylspermine. This in turn is degraded by polyamine oxidase with the formation of spermidine and an aldehyde, 3-acetamidopropanal (Fig. 9). Spermidine also is acetylated by the same transferase and the $N^1$-acetylspermidine so formed can be cleaved by polyamine oxidase to form putrescine and acetamidopropanal. The putrescine can then be recycled (Fig. 4) or further metabolized to $\gamma$-aminobutyric acid (Fig. 10), and $\beta$-alanine (3-aminopropionic acid) can be derived from the aldehyde (Fig. 11).[25]

### III. Regulation of Polyamine Biosynthesis

The three key enzymes that regulate the pathway shown in Fig. 4 are ornithine decarboxylase, $S$-adenosylmethionine decarboxylase and spermidine/spermine $N^1$-acetyltransferase;[26] the activities of the other enzymes appear to be governed primarily by the availability of the appropriate substrate.

### A. ORNITHINE DECARBOXYLASE

Ornithine decarboxylase has been the most studied of the enzymes in the polyamine biosynthetic pathway. The mammalian enzyme has proved difficult to purify as it is extremely labile, the cellular content is very low, and it may be present in multiple forms. Ornithine decarboxylase has been isolated from a number of microorganisms, from rat liver, calf liver and most recently from kidneys of androgen-treated mice.[27] The latter enzyme was found to have a $M_r$ of 53 000 by gel electrophoresis in the presence of sodium dodecyl-sulphate, and $M_r$ 100 000 by gel filtration, indicating that it occurs as a dimer. The $K_m$ for ornithine was 75 $\mu$M. Ornithine decarboxylase is pyridoxal-requiring and the $K_m$ of the mouse kidney enzyme for pyridoxal 5'-phosphate is 0.3 $\mu$M. The enzyme occurs in both nucleus and cytoplasm, but after induction most of the activity is cytoplasmic. The recent production of monospecific, high affinity antibodies to this enzyme,[28–31] together with the availability of the irreversible enzyme-activated, specific inhibitor difluoromethylornithine (Fig. 12),[32] have led to the cloning of the ornithine

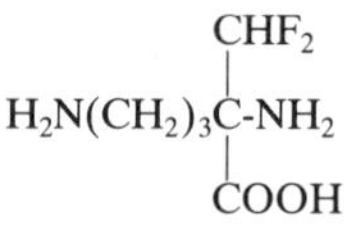

$$\overset{\displaystyle CHF_2}{\underset{\displaystyle COOH}{H_2N(CH_2)_3\overset{|}{\underset{|}{C}}\text{-}NH_2}}$$

Fig. 12. Structure of difluoromethylornithine.

decarboxylase gene from mouse kidney[33–35] and the deduction of its amino acid sequence from the nucleotide sequence of the complementary DNA.[36,37] The calculated $M_r$ based on the amino acid composition was 51 172.

Eukaryotic cells and *E. coli* contain proteins which are induced by di- and polyamines that non-competitively inhibit ornithine decarboxylase and to which the name antizyme has been applied.[38] In addition, mammalian cells contain an anti-antizyme, a protein that specifically binds to the antizyme of an ornithine-antizyme complex and liberates free enzyme. The role of these inhibitory proteins in the control of ornithine decarboxylase activity is by no means clear. Interestingly, the antizyme from *E. coli* will inhibit the rat liver enzyme and vice-versa, and this together with the demonstration that antisera to the mouse kidney or rat liver decarboxylase will cross-react with ornithine decarboxylase from all mammalian species examined so far, indicate that these aspects óf the control of polyamine biosynthesis have been highly conserved during evolution.

## B. S-ADENOSYLMETHIONINE DECARBOXYLASE

This, the second of the rate-limiting enzymes in polyamine biosynthesis, has been purified from *E. coli*, *Saccharomyces cerevisiae*, and from rat liver.[39] In each case the enzyme contains covalently-bound pyruvate, an unusual feature in a eukaryote enzyme. A summary of the properties of these three enzymes is given in Table 2. All are inhibited by their product, decarboxylated S-adenosylmethionine (Fig. 7), and by methylglyoxal *bis*-(guanylhydrazone), a structural analog of spermidine (Fig. 13) that also has other effects on polyamine metabolism including acting as an inhibitor of polyamine oxidase[40] and an inducer of spermidine/spermine $N^1$-acetyl-transferase.[11] In contrast to ornithine decarboxylase, no specific inhibitor for this enzyme is available. The rat liver and the yeast decarboxylases are both activated by putrescine (Table 2) but not by divalent cations, whereas the opposite is true of the bacterial and plant enzymes.[17] Evidence for the existence of multiple forms of this enzyme is provided by the demonstration that S-adenosylmethionine decarboxylase from rat psorias muscle differs from the liver enzyme in $K_m$ (100 $\mu$M as opposed to 50 $\mu$M), sensitivity to stimulation by putrescine (15-fold compared to 6-fold), and to inhibition by spermidine.[26,41]

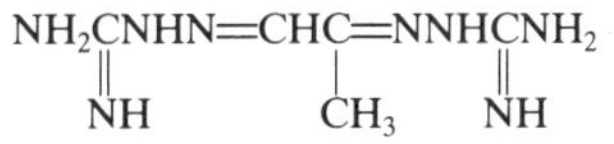

$$NH_2CNHN{=}CHC{=}NNHCNH_2$$
$$\underset{NH}{\|} \qquad \underset{CH_3}{|} \quad \underset{NH}{\|}$$

**Fig. 13.** Methylglyoxal *bis*(guanylhydrazone).

TABLE 2

Properties of $S$-adenosylmethionine isolated from three different sources

| Source | $M_r$ | Subunit $M_r$ (and No.) | Activators | $K_m$ ($\mu$M) |
|---|---|---|---|---|
| E. coli | 108 000 | 17 000 (6) | $Mg^{++}, Ca^{++}, Mn^{++}$ | 60 |
| S. cerevisiae | 88 000 | 41 000 (2) | Putrescine | 151 |
| Rat liver | 68 000 | 32 500 (2) | Putrescine | 50 |

Data from ref. 39.

## C. SPERMIDINE/SPERMINE $N^1$-ACETYLTRANSFERASE

Two polyamine acetylases are known. One is a predominantly nuclear enzyme[42] that acts mainly on histones but can acetylate spermidine to form $N^8$-acetylspermidine, which is then deacetylated within the cell. The purpose of this procedure is not clear but, by analogy with what is known of the effects of acetylation on the cellular uptake of polyamines (p. 104), may provide a means of removing unwanted spermidine from within the nucleus. The second enzyme, spermidine/spermine $N^1$-acetyltransferase, is cytosolic and can be distinguished from the nuclear acetylase by reaction with specific antibodies, and its differing response to certain inhibitors.[43] The enzyme has been purified from rat liver[44] and has an apparent $M_r$ of about 115 000 made up of two subunits of 60 000. It acts on spermidine to form exclusively $N^1$-acetylspermidine. Spermine is acetylated to give $N^1$-acetylspermine which can be further acetylated by this enzyme to $N^1,N^{12}$-diacetylspermine (Fig. 14), although this compound has not been detected *in vivo*. The Michaelis constants for spermidine, spermine, and $N^1$-acetylspermine were

$$CH_3CONH(CH_2)_4NH_2$$
$N^1$-acetylputrescine
$$CH_3CONH(CH_2)_3NH(CH_2)_4NH_2$$
$N^1$-acetylspermidine
$$NH_2(CH_2)_3NH(CH_2)_4NHCOCH_3$$
$N^8$-acetylspermidine
$$CH_3CONH(CH_2)_3NH(CH_2)_4NH(CH_2)_3NH_2$$
$N^1$-acetylspermine
$$CH_3CONH(CH_2)_3NH(CH_2)_4NH(CH_2)_3NHCOCH_3$$
$N^1,N^{12}$-diacetylspermine

**Fig. 14.** Acetylated derivatives of putrescine, spermidine and spermine. The terminology used here for the $N$-substituted derivatives of spermine and spermidine is that of Tabor,[191] the numbering being such that the secondary amino group has the lowest possible number.

130 $\mu$M, 35 $\mu$M, and 30 $\mu$M, respectively. The enzyme appears to be highly specific for an aminopropyl moiety attached to a secondary amino group,[45] i.e. substrates of the type $H_2N(CH_2)_3NHR$. The purified enzyme has no deactylase activity, hence the reaction is essentially irreversible.

These three key enzymes are characterized by their extremely short half-lives (of the order of 1 h or less), the ease with which each can be induced by a variety of agents, and their presence in unstimulated cells only in very small amounts (Pegg[11] has calculated that ornithine decarboxylase represents about 1 part in $6 \times 10^7$ of the soluble protein in normal rat liver). Increase in the activity of these enzymes is dependent both on the synthesis of new enzyme protein and a reduction in the rate at which the enzyme is degraded.

Induction of ornithine decarboxylase appears to be a universal accompaniment to the stimulation of cell growth by a vast number of hormones, drugs, tumour promotors and other stimuli.[46] The activity of the enzyme is reduced in the presence of putrescine and other polyamines.

Application of exogenous polyamines also results in a reduction in the activity of $S$-adenosylmethionine decarboxylase.[39] Conversely, polyamine depletion leads to an increase in the synthesis of this enzyme, which is activated by putrescine (Table 2).

Synthesis of spermidine/spermine $N^1$-acetyltransferase can be induced by a wide variety of toxic stimuli, hormones, exogenous polyamines and some synthetic analogues, and serum growth factors.[11,47–49]

## IV. Polyamine Degradation

In contrast to the extensive studies of polyamine biosynthesis, polyamine degradation has received much less attention, although the *in vitro* cytotoxic or cytostatic effects of enzymically-oxidized polyamines have been well documented in a wide variety of cell types.[50–52]

## A. NOMENCLATURE OF THE POLYAMINE OXIDASES

The nomenclature of the polyamine oxidases is confused.[11,52–54] The current Enzyme Commission Handbook[18] lists two amine oxidases: EC 1.4.3.4 amine oxidase (flavin-containing) and EC 1.4.3.6 amine oxidase (copper-containing). An alternative, and perhaps more logical grouping would be to divide the polyamine oxidases into those that act at the primary amino groups, and those (the majority) that act at the secondary amino group(s) of the aminopropyl moieties of spermine or spermidine. Those

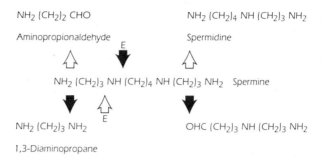

**Fig. 15.** Alternative sites of cleavage of spermine by polyamine oxidases from various sources. The substrate is cleaved at a secondary amino group with the formation of aminopropionaldehyde or diaminopropane as one of the products. Spermidine is attacked in a similar manner.

enzymes that act at the secondary amino group would be further subdivided according to whether diaminopropane or aminopropionaldehyde were among the products (Fig. 15). For the purposes of this essay amine oxidases able to utilize the polyamines spermine or spermidine as substrates will be considered as polyamine oxidases, whether or not they can also act on mono- or di-amines.

## B. BOVINE PLASMA POLYAMINE OXIDASE

In the 30 years since its discovery the bovine plasma polyamine oxidase (EC 1.4.3.6) has been the subject of extensive study. It has been purified to apparent homogeneity and crystallized.[55] However, electrophoresis in polyacrylamide disc gels, with or without sodium dodecyl sulphate, revealed the presence of minor (<5%) components that also possess amine-oxidizing activity[56,57] and appear to share a common mechanism.[58]

The enzyme acts on spermidine and spermine to produce respectively an aminomonoaldehyde [$N'$-(4-aminobutyl)-aminopropionaldehyde] or a dialdehyde [$NN'$-*bis*(3-propionaldehyde)-1,4-diaminobutane], ammonia and hydrogen peroxide (Fig. 16).[59] The aminoaldehyde products cannot be recycled (Fig. 4), but must enter some terminal pathway of polyamine catabolism. The enzyme, $M_r$ about 180000,[57,60] consists of two non-covalently linked subunits of $M_r$ 90000 each. Association–dissociation between monomer–dimer–trimer forms has been noted,[60] providing an explanation for reported molecular weights of the order of 265000.[61–63] Each subunit contains a disulphide bridge, and there are two copper atoms and one carbonyl group per molecule. Only one of the copper atoms appears to be essential for activity,[64] as is the carbonyl group. Early reports suggested

$$NH_2(CH_2)_3NH(CH_2)_4NH(CH_2)_3NH_2 + 2O_2 + 2H_2O \rightarrow$$
spermine

$$OCH(CH_2)_2NH(CH_2)_4NH(CH_2)_2CHO + 2NH_3 + 2H_2O_2$$
$N,N'$-bis(3-propanal)-1,4-diaminobutane

$$NH_2(CH_2)_4NH(CH_2)_3NH_2 + O_2 + H_2O \rightarrow$$
spermidine

$$NH_2(CH_2)_4NH(CH_2)_2CHO + NH_3 + H_2O_2$$
$N$-(4-aminobutyl)-3-aminopropanal

Fig. 16. Oxidation of spermine and spermidine by bovine plasma amine oxidase.

pyridoxal phosphate as the carbonyl containing cofactor[65] but later work has shown that this is not so,[66] and it has recently been reported that the prosthetic group is covalently bound pyroloquinoline quinone (Fig. 17),[67] a cofactor for a number of dehydrogenases.[68]

The enzyme is a glycoprotein, containing 7–8% carbohydrate attached by a glycosylamine linkage between $N$-acetylglucosamine and an asparagine residue in the protein.[69] The carbohydrate moiety is not essential for activity and is remote from the active site.[70] Three forms of the enzyme have been reported[55,56–58] but data on the minor components is sparse. Yasunobu and co-workers have suggested[56] that these multiple forms may be isoenzymes that differ only in carbohydrate content, but this has yet to be established.

The enzyme oxidizes polyamines containing primary amino groups, those forming part of an aminopropyl moiety being the more readily attacked,[71] and some primary amines including benzylamine.[55,72] Molecular oxygen is required, as spermine was not degraded by the enzyme in a helium atmosphere.[72] Cadaverine (1,5-diaminopentane), [lysine]-vasopressin, the peptide Pro-Lys-Gly-NH$_2$, and elastin are also oxidized slowly.[62] The enzyme is

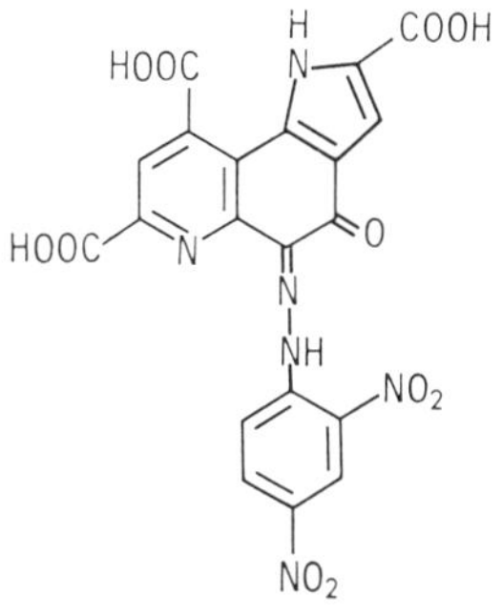

Fig. 17. Pyroloquioline quinone.

inhibited by carbonyl[72] and sulphhydryl reagents,[73] copper chelators,[74] aminoguanidine and methylglyoxal *bis*(guanylhydrazone),[63] and β-amino-propionitrile,[62] which also inhibits lysyl oxidase,[75,76] an enzyme which may bear more than a passing resemblance to the bovine plasma amine oxidases.[77]

After some 30 or more years investigation of the bovine plasma enzyme by many laboratories it is salutary to list what we do not yet know about it. We do not know the amino acid composition or sequence, although a preliminary composition of one form was published in 1976.[56] Neither do we know the mechanism by which the enzyme oxidizes spermine or spermidine, preferentially attacking the aminopropyl end—we are not even certain if these are the natural substrates of this enzyme. We have no information on the rate of secretion of the enzyme, its half-life in the circulation, tissue of origin, or its function. Nor do we know why similar enzymes have been found in the sera of all other ruminants examined,[78] and in some connective tissues,[52,62] but not, apparently, in sera from non-ruminants, although Seiler and co-workers[79,80] have presented indirect evidence for the presence in murine sera and tissues of such an enzyme following the intraperitoneal administration of large doses of spermine or spermidine. The demonstration of N-(3-aminopropyl)pyrrolidin-2-one[81] and a variety of amino acids derived from spermine or spermidine (Fig. 18) in normal human and rat urine[82–85]

I. $NH_2(CH_2)_3COOH$
   4-amino-*n*-butyric acid
   γ-aminobutyric acid

II. $HOOC(CH_2)_2NH(CH_2)_3COOH$
    N-(2-carboxyethyl)-4-amino-*n*-butyric acid

$HOOC(CH_2)_2NH(CH_2)_4NH_2$
N-2(2-carboxyethyl)-1,4-diaminobutane
N-(4-aminobutyl)-3-aminopropionic acid
putreanine

$NH_2(CH_2)_3NH(CH_2)_3COOH$
N-(3-carboxypropyl)-1,3-diaminopropane
(isoputreanine)

III. $NH_2(CH_2)_3NH(CH_2)_4NH(CH_2)_2COOH$
     N-(3-aminopropyl)-N'-(2-carboxyethyl)-1,4-diaminobutane
     $N^8$-(2-carboxyethyl)spermidine

$HOOC(CH_2)_2NH(CH_2)_4NH(CH_2)_2COOH$
N,N'-bis(2-carboxyethyl)-1,4-diaminobutane
(spermic acid)

**Fig. 18.** Amino acids derived from putrescine (I), spermidine (II) and spermine (III). All have been found in normal human and rat urine.

has also been interpreted as indicating the presence in these species of an enzyme able to oxidatively deaminate the terminal aminopropyl groups of spermine and spermidine. Human seminal plasma may contain a similar enzyme.[86]

## C. TISSUE POLYAMINE OXIDASE

Holtta[87,88] isolated and characterized an enzyme from rat liver that catalysed the oxidation of spermine, spermidine and derivatives formed by the acetylation of their aminopropylmoieties; benzylamine was not oxidized. The highest specific activity was found in the peroxisomal fraction, where its presence has been confirmed histochemically in both rat liver and kidney.[89] The enzyme is a single polypeptide of $M_r$ 60 000, containing tightly bound FAD as the prosthetic group, and possibly iron.[88] It acts on the secondary amino groups of spermine or spermidine with the production of 3-amino-propanal as shown in Fig. 19. This is the FAD-dependent enzyme of Fig. 4, responsible for the recycling of spermine and spermidine to putrescine.

Sulphhydryl and carbonyl reagents inhibited the enzyme, whose activity was increased in the presence of dithiothreitol or mercaptoethanol.[87] Thus the enzyme appears to contain both sulphhydryl and carbonyl groups that are essential for activity. Typical inhibitors of pyridoxal phosphate requiring enzymes, or copper chelators were without effect. Iron chelators were inhibitors. Quinacrine ($N^4$-(6-chloro-2-methoxy-9-acridinyl)-$N'$,$N'$-diethyl-1,4-pentanediamine; 6-chloro-9[(4-diethylamino)-1-methyl-butyl]-amino-2-methoxyacridine), a flavoprotein inhibitor, which does not inhibit bovine plasma amine oxidase, strongly inhibited the rat liver enzyme. Oxygen appears to be the sole electron acceptor. Apparent Michaelis constants of 20 $\mu$M and 50 $\mu$M were obtained for spermine and spermidine, respectively.[87] $N^1$-Acetylspermidine ($K_m$ 14 $\mu$M), $N^1$-acetylspermine (0.6 $\mu$M), and $N^1$,$N^{12}$-diacetylspermine (5 $\mu$M) are also substrates for this enzyme.[90]

$$NH_2(CH_2)_3NH(CH_2)_4NH(CH_2)_3NH_2 + O_2 + H_2O \rightarrow$$
$$\text{spermine}$$

$$NH_2(CH_2)_3NH(CH_2)_4NH_2 + NH_2(CH_2)_2CHO + H_2O_2$$
$$\text{spermidine} \qquad \text{3-aminopropanal}$$

$$NH_2(CH_2)_3NH(CH_2)_4NH_2 + O_2 + H_2O \rightarrow$$
$$\text{spermidine}$$

$$NH_2(CH_2)_2CHO + NH_2(CH_2)_4NH_2 + H_2O_2$$
$$\text{3-aminopropanal} \qquad \text{putrescine}$$

**Fig. 19.** Oxidation of spermine and spermidine by rat liver polyamine oxidase.

TABLE 3

Relative distribution of polyamine oxidase in mammalian tissues

| | Rat | Human[a] | Human |
|---|---|---|---|
| Pancreas | 38·1 | 1 | — |
| Liver | 20·2 | 25·0 | — |
| Spleen | 15·2 | 8·8 | — |
| Kidney | 14·9 | 16·5 | — |
| Small intestine | 10·8 | 2·6 | — |
| Testes | 9·8 | 21·3 | — |
| Thymus | 9·3 | — | — |
| Prostate (ventral) | 6·9 | — | — |
| Brain | 6·3 | 2·2 | — |
| Lung | 4·3 | 1·9 | — |
| Heart | 2·0 | 2·4 | — |
| Skeletal muscle | 1·0 | — | — |
| Placenta | — | — | 1·0 |
| Decidua | — | — | 9·1 |
| Amnion | — | — | 5·7 |
| Chorion | — | — | 3·1 |
| Substrate | $N^1,N^{12}$-diacetyl -spermine | $N^1$-acetyl spermine | [$^{14}$C]-spermine |
| Assay | TLC of dansyl derivatives | $H_2O_2$ production | [$^{14}$C]- spermidine |
| Reference | 91 | 94 | 92, 93 |

[a]Cadaver tissues.

Polyamine oxidase activity is high in most tissues. It is present at levels comparable to those of spermine and spermidine synthase and that greatly exceed that of spermine/spermidine $N^1$-acetyltransferase.[11] Hence tissue levels of acetylated polyamines verge on the undetectable.[25] The enzyme is present in all mammalian tissues and in all cultured mammalian cells examined so far. However, data on the distribution of the enzyme in tissues is available for only two species, the rat[91] and man (Table 3).[92-94] Relative activities have been calculated from each set of data, as the use of different substrates and assay conditions makes direct comparison of the results exceedingly difficult.

## D. HUMAN PREGNANCY-ASSOCIATED POLYAMINE OXIDASE

An association between maternal blood amine oxidase activity and pregnancy has been recognized for more than 40 years.[40,54,95,96] Histaminase,[97]

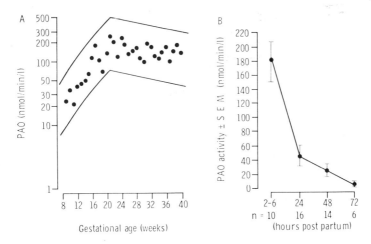

**Fig. 20.** Human serum polyamine oxidase activity. (A) In normal pregnancy, points are observed means, solid lines are 95% reference limits ($n = 240$).[101] (B) In the early post-partum period.[92]

diamine (putrescine) oxidase,[98] spermidine oxidase[99] and polyamine (spermine) oxidase,[100,101] which may or may not all be the same enzyme, show a progressive increase in activity with increase in gestational age to 21 weeks or beyond, and decline to vanishingly low levels 3–4 days after delivery (Fig. 20). Only traces of similar activities are found in male sera or in sera from non-pregnant women of reproductive age. The placenta has long been regarded as a source of these pregnancy-associated enzymes but there is evidence that they may also originated from the decidua.[92,102,103]

Smith[104] partially purified (500-fold) an enzyme from human placenta that oxidized cadaverine ($K_m$ 0·2 $\mu$M), putrescine (0·07 $\mu$M), diaminopropane, histamine and benzylamine. Paolucci and co-workers[105] obtained a placental amine oxidase that on gel filtration gave four active fractions with molecular masses that were multiples of 125 000 ± 5000. The enzyme oxidized putrescine ($K_m$ 330 $\mu$M), histamine (6 $\mu$M), cadaverine (3·3 $\mu$M), diaminopropane and spermine.

The most extensive study of placental amine oxidases is that of Bardsley and co-workers[106–110] who isolated two amine oxidases from homogenates of human placenta. One was a single polypeptide of $M_r$ 70 000 containing 1 g.atom each of copper and manganese,[110] that oxidized a large number of amines including o-*bis*(aminomethyl)benzene, cadaverine, putrescine, spermine and histamine in the ratio 214 : 147 : 100 : 32 : 30, but not spermidine or benzylamine. [8-arginine]Vasopressin, [8-lysine]vasopressin ($k_m$ 18 $\mu$M), collagen and tropocollagen were also oxidized. The second enzyme, which

$$CH_3CONH(CH_2)_3NH(CH_2)_4NH(CH_2)_3NHOCH_3 + O_2 + H_2O \rightarrow$$
$$N^1,N^{12}\text{-diacetylspermine}$$
$$CH_3CONH(CH_2)_3NH(CH_2)_4NH_2 + OCH(CH_2)_2NHOCH_3 + H_2O_2$$
$$N^1\text{-acetylspermidine} \qquad 3\text{-acetamidopropanal}$$

Fig. 21. Oxidation of $N^1,N^{12}$-diacetylspermine by human polyamine oxidase.

was not inhibited by aminoguanidine, also oxidized a number of substrates including adrenaline (relative rate 1300), putrescine (100), cadaverine (84), spermine (50), spermidine (140), histamine (49) and benzylamine (67).

Gahl and co-workers[99] were unable to separate the putrescine- and spermidine-oxidizing activities in human pregnancy serum and obtained $K_m$ values of 2·5 $\mu$M and 10·9 $\mu$M, respectively. Both activities were competitively inhibited by aminoguanidine.

A polyamine oxidase of $M_r$ about 67 000 has been partially purified[40,111] from human retroplacental blood serum, known to be a particularly rich source of this enzyme.[100] Oxidation of spermine and spermidine resulted in the production of aldehydes in the molar ratios of 2:1 and 1:1 respectively, but neither ammonia nor superoxide formation could be detected. The reaction requires molecular oxygen as anoxic conditions resulted in a decrease in enzyme activity. Oxidation of $N^1,N^{12}$-diacetylspermine gave $N^1$-acetylspermidine, in accordance with the scheme in Fig. 21. Thus the mode of action of this enzyme resembles that of the rat liver, or tissue, polyamine oxidase and differs from that of the bovine plasma amine oxidase. The enzyme was completely inhibited by Quinacrine and to a lesser extent by $\beta$-aminoproprionitrile, aminoguanidine and methylglyoxal bis(guanylhydrazone).[40,111] Polyamine oxidase activity has recently been discovered in human milk.[112]

More than 40 years of investigation have yielded only meagre knowledge of the function(s) of the amine oxidases in pregnancy. It seems probable that in human tissues there are, to quote Crabbe,[113] 'a range of catalytic proteins with amine oxidising activity, some preferring monoamines to diamines, some vice-versa, and some polyamines or protein-bound lysyl groups'. In pregnancy one or more of these enzymes is released by the placenta, and possibly the decidua, and appears in the circulation in quantities that, when compared with the tissue levels and half-life of the enzyme,[92] suggest active secretion. Plasma spermine is also increased in pregnancy.[114,115] In a preliminary study Illei and Morgan,[116] found that serum polyamine oxidase activity was significantly lower in spontaneous abortion compared to normal pregnancies of the same gestational age. An immunosuppressive role has been proposed,[117-123] based on observations that cells involved in the immune response appear to be more susceptible to the effects of oxidized polyamines than were other cells such as fibroblasts. However, this may be merely a special adaptation of a much more general function.

## E. POLYAMINE OXIDASES OF MICROORGANISMS AND PLANTS

As might be expected, plants and bacteria can do things differently. Polyamine oxidase activity has been found in fungal mycelia[124,125] and the enzyme has been purified and crystallized from extracts of *Penicillium chrysogenum* or *Aspergillus terreus*.[126–128] Both enzymes are of the rat liver type. A peroxisomal polyamine oxidase from the yeast *Candida Boidinii* oxidized di-*n*-butylamine to butyraldehyde and hydrogen peroxide.[129] Spermidine was oxidized to putrescine and 3-aminopropanal. The amine oxidase from *Pichia pastoris*[130] resembles the bovine plasma enzyme in its action on spermine, but for spermidine, the reaction postulated (Fig. 22) differs from that for any other amine oxidase so far described.

$$NH_2(CH_2)_3NH(CH_2)_4NH_2 + 2O_2 + 2H_2O \rightarrow$$
$$OCH(CH_2)_2NH(CH_2)_3CHO + 2NH_3 + 2H_2O$$

**Fig. 22.** Oxidation of spermidine by *Pichia pastoris* amine oxidase.

Copper- and carbonyl-containing enzymes capable of oxidizing a range of primary amines containing the —$CH_2$—$NH_2$ group are widespread in the Leguminosae.[17,131,132] Spermidine is oxidized to a compound that was thought to be 1-(3-aminopropyl)-1-pyrroline, and putrescine to pyrroline (Fig. 23); ammonia and hydrogen peroxide are also produced. The structure of the pyrroline-like product has recently been shown to be the *N,N'*-acetal, octahydropyrrol[1,2-*a*]pyrimidine, called also 1,5-diazabicyclo[4.3.0]-nonane; Fig. 24.[133,134] The Gramineae contain an oxidase specific for polyamines that acts on spermine or spermidine to form diaminopropane and a similar pyrroline-like compound (Fig. 25), but not ammonia.[17] The maize seedling enzyme has been shown to contain FAD,[135] but this has not been detected with certainty in oxidases from other cereals. Similar FAD-

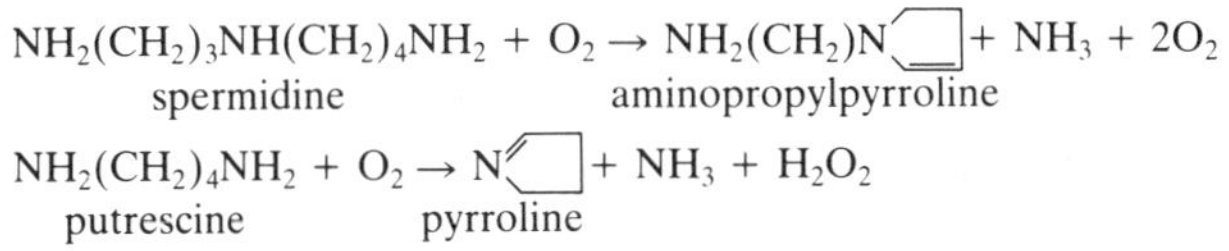

$$NH_2(CH_2)_3NH(CH_2)_4NH_2 + O_2 \rightarrow NH_2(CH_2)N\underline{\phantom{xx}} + NH_3 + 2O_2$$
spermidine                          aminopropylpyrroline

$$NH_2(CH_2)_4NH_2 + O_2 \rightarrow N\underline{\phantom{xx}} + NH_3 + H_2O_2$$
putrescine                 pyrroline

**Fig. 23.** Oxidation of putrescine and spermidine by the amine oxidase in Leguminosae.

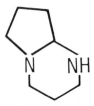

**Fig. 24.** Structure of 1,5-diazabicyclo[4.3.0]nonane.

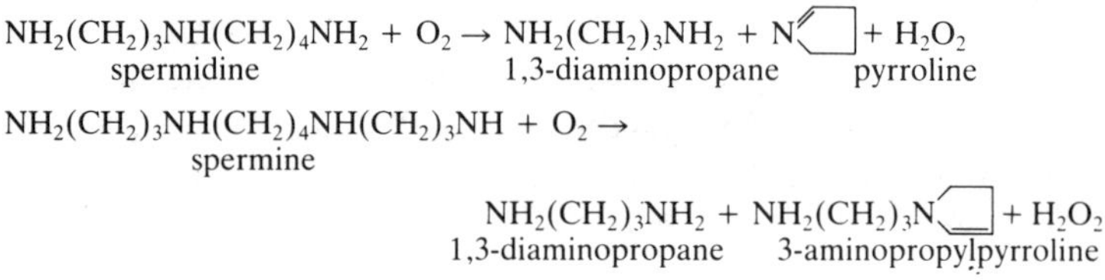

$$NH_2(CH_2)_3NH(CH_2)_4NH_2 + O_2 \rightarrow NH_2(CH_2)_3NH_2 + N\boxed{\phantom{x}} + H_2O_2$$
$$\text{spermidine} \qquad\qquad\qquad \text{1,3-diaminopropane} \quad \text{pyrroline}$$

$$NH_2(CH_2)_3NH(CH_2)_4NH(CH_2)_3NH + O_2 \rightarrow$$
$$\text{spermine}$$

$$NH_2(CH_2)_3NH_2 + NH_2(CH_2)_3N\boxed{\phantom{x}} + H_2O_2$$
$$\text{1,3-diaminopropane} \quad \text{3-aminopropylpyrroline}$$

**Fig. 25.** Oxidation of spermidine and spermine by polyamine oxidase from the Graminae.

containing enzymes are present in bacteria such as *Serratia, Micrococcus,* and *Pseudomonas*.[52,125]

## V. Perspectives

The ubiquitous occurrence of polyamines in living organisms has prompted the suggestion that to ask "what do the polyamines do?" is like asking "what is the role of potassium or magnesium?" Indeed, Cohen[136] goes on to point out that "the notion that biochemical substances have only a single role is dangerous to investigators and unlikely to be correct".

The early suggestions that the polyamines were merely organic cations that could variably substitute for calcium or magnesium are now disproved. However, it is possible arbitrarily to divide the known roles of the polyamines into those that relate to their ability to provide flexible and extensible, sterically spaced, cationic centres (i.e. that depend on physico-chemical properties), and those that follow from "biological" properties. Examples of the former are the interactions of spermidine and spermine with nucleic acids and their effects on transcription and translation.[137-139] However, the specificity and mechanism of polyamine interaction with nucleic acids is still obscure.

With regard to the second category we still do not know the intracellular functions of the polyamines at the molecular level, although it is clear that they are essential for cell growth and differentiation.[11,140] Is ornithine decarboxylase the easily induced, labile protein, the synthesis of which has been postulated[141] to be a critical requirement for cells to proceed beyond the restriction point in the cell cycle? Why do cells possess a saturable, energy-requiring transport system for polyamines[142,143] which can also take up structural analogues such as methylglyoxal *bis* (guanylhydrazone),[144] and paraquat?[145] and why do some proteins contain covalently bound polyamines?[146] Scalabrino and Ferioli[147] have suggested that the polyamines and their oxidized derivatives form part of an integrated biochemical system for the regulation of cell proliferation and growth, by means of stimulatory substances (the polyamines) and inhibitory substances (the oxidized polyamines or derivatives), but the evidence is far from complete.

The secondary or terminal routes of polyamine metabolism are also largely unknown. Several attempts have been made to construct comprehensive pathways of polyamine metabolism.[79,148,149] A difficulty with this approach lies in the need to include oxidases that act on primary amino groups and those that act on secondary amino groups (although one may be intracellular and another extracellular) in order to account for all the products known to be excreted in urine. It is clear that the aminoaldehydes resulting from polyamine oxidation are cytotoxic,[50–52,150,151] but the mechanism and *in vivo* relevance of this effect is unknown. Are the reported similarities between some polyamine oxidases and lysyl oxidase[77] more apparent than real? This question also awaits an unequivocal answer.

The published values for molecular weights of chalones (the tissue-specific, species-non-specific, regulators of tissue growth) have steadily decreased over the years.[152,153] This, and other evidence indicating a possible polyamine involvement, led to the suggestion[52,154] that chalones may be oxidized polyamines that readily, and reversibly, bind to proteins by virtue of possessing both positively-charged amino groups and a carbonyl group in a relatively small molecule. However, this is an area of research beset with pitfalls, and in an extensive review of this field Rytomaa[155] observed that "experimental evidence does not seem to keep pace with theoretical considerations in chalone research".

Further evidence for the possible involvement of polyamine oxidases, or oxidized polyamines, in processes regulating cell division is provided by the work of Quash and co-workers[156] who found that polyamine oxidase activity was three- to five-fold lower in transformed than in normal cells, and the finding of elevated levels of $N^1$-acetylspermidine in human breast cancers.[157]

What is the role of the polyamine oxidases in cellular function? The widespread occurrence and similarities (see Table 4 for examples) of the polyamine-cleaving oxidases (Fig. 15) in higher plants,[17] bacteria, fungi,

TABLE 4

Similarities in the properties of polyamine oxidases from various sources

| Source | MW values determined by gel filtration | | | Products$^a$ include | Refs |
|---|---|---|---|---|---|
| | MW | Subunits | FAD | | |
| *Penicillium chrysogenum* | 160 000 | 2 | + | Aminepropionaldehyde | 126 |
| *Aspergillus terreus* | 130 000 | 2 | + | Aminepropionaldehyde | 127 |
| *Micrococcus rubens* | 80 000 | 1 | + | Diaminopropane | 194 |
| *Zea mays* | 65 000 | 1 | + | Diaminopropane | 135 |
| *Avena sativa* | 85 000 | 1 | ? | Diaminopropane | 131 |
| Rat liver | 60 000 | 1 | + | Aminopropionaldehyde | 87 |
| Human pregnancy serum | 67 000 | 1(?) | ? | Aminopropionaldehyde | 111 |

$^a$Spermine or spermidine as substrate.

protozoa,[158] macrophages,[159] lymphocytes,[160] fibroblasts, vascular endothelial and smooth muscle cells, human trophoblasts (D. M. L. Morgan, unpublished data) and all mammalian tissues examined (Table 3) indicate that these enzymes have been well conserved throughout evolution. This in turn suggests that the enzymes or their products play a fundamental role in cell function. However, as has already been mentioned, the aminoaldehyde products of polyamine oxidation are cytotoxic.[50–52,150,151] The discovery that acetylation of the aminopropyl groups prior to oxidation prevents this cyctotoxicity[150,161,162] (perhaps by reducing the charge on the amino nitrogen) may provide the answer to a puzzling question: why should polyamine degradation apparently proceed by a pathway that results in the production of such potentially hazardous compounds as the aminoaldehydes? The reported kinetic constants of rat liver polyamine oxidase for $N^1$-acetylspermine, $N^1,N^{12}$-diacetylspermine and spermine suggest that the acetyl derivatives will be metabolized more rapidly than the parent compound. The spermidine formed by oxidation of $N^1$-acetylspermine would be acetylated in turn to form $N^1$-acetylspermidine (the acetyltransferase can be induced by spermidine),[163] oxidation of which would also result in products that were not cytotoxic.

Acetylation of the primary amino groups prevented the uptake of polyamines by cultured L1210 leukaemia cells[164,165] and this, together with the occurrence of acetylated polyamines in normal urine[148,166] suggests that acetylation may be an important step in polyamine elimination[167] as well as in degradation via the polyamine oxidase pathway.[149] Indeed, it has been suggested[168] that acetylation may be a means whereby biologically-active polyamines may be inactivated, and this may explain the role of the nuclear acetylase described earlier. Two new enzyme-activated, irreversible inhibitors, $N^1$-methyl-$N^2$-(2,3-butadienyl)-1,4-butanediamine and $N^1,N^2$-bis(2,3-butadienyl)-1,4-butanediamine,[169] which appear to be specific for polyamine oxidase, have been used to explore the effects of inhibition of this enzyme on polyamine metabolism in vivo.[170,172] In each case there was an increased accumulation and/or excretion of acetyl polyamines following treatment with the inhibitor. These compounds may prove to be powerful tools with which to investigate polyamine degradation, both in vitro and in vivo.

The enhanced polyamine metabolism that takes place in rapidly growing tissues has prompted considerable interest in the use of polyamines as tumour markers or as indices of therapy.[114,173] By and large, this approach has proved less successful than expected,[174] due in part to methodological problems,[175] and it is now clear that measurements of urinary polyamines do not provide a specific indication of malignant disease. However, recently it has been shown that in heart transplant patients changes in urinary polyamines and their acetylated derivatives predict rejection episodes prior to any other clinical test known to date.[176] Another potentially fruitful area is the use of analogues of the polyamines or inhibitors of polyamine biosynthesis, in cancer chemotherapy.[164,165,177]

The discovery that the oxidation products of spermine or spermidine can kill intraerythrocytic parasites without lysing the host red cell[178] stimulated interest in this system.[179–184] The ability to interfere in parasite polyamine metabolism in this way, or by the therapeutic administration of inhibitors of biosynthesis, offers exciting possibilities for the treatment of diseases that are endemic in many areas.[11,185,188]

## VI. Conclusions

There is still much to learn about the role of the polyamines in cell growth and proliferation. Their ubiquity, the potency of some of their metabolites, and the evolutionary conservation of some of the enzymes of polyamine metabolism, indicates that their role(s) are important. It is clear that polyamine biochemistry, which has spread throughout the biomedical sciences, offers exciting prospects for the future.

## ACKNOWLEDGEMENTS

I thank Elsie Prestige for her help in the preparation of this manuscript.

# REFERENCES

1. Leuwenhoek, A. van (1678). Observationes D Anthonii Leuwenhoek de natis e semine genitali animalculis. *Phil. Trans.* **12**, 1040–1043.
2. Bachrach, U. (1973). *Function of the Naturally Occurring Polyamines.* Academic Press, New York and London.
3. Dudley, H. W., Rosenheim, O. & Starling, W. W. (1926). The chemical constitution of spermine. III. Structure and synthesis. *Biochem. J.* **20**, 1082–1094.
4. Dudley, H. W., Rosenheim, O. & Starling, W. W. (1927). The constitution and synthesis of spermidine, a newly discovered base isolated from animal tissues. *Biochem. J.* **21**, 97–103.
5. Brieger, I. (1879). Uber die aromatischen Produkte der Faulniss aus Eiweiss. *Z. Physiol. Chem.* **iii**, 134–148.
6. Landenburg, A. (1886). Uber Pentamethylendiamin und Tetramethylendiamin. *Ber.* **19**, 780–783.
7. Cohen, S. S. (1971). *Introduction to the Polyamines.* Prentice-Hall, New Jersey.
8. Raina, A. & Janne, J. (1975). Physiology of the natural polyamines putrescine, spermidine and spermine. *Med. Biol.* **53**, 121–147.
9. Tabor, C. W. & Tabor, H. (1976). 4-Diaminobutane (putrescine), spermidine, and spermine. *Ann. Rev. Biochem.* **45**, 285–306.
10. Janne, J., Poso, H. & Raina, A. (1978). Polyamines in rapid growth and cancer. *Biochim. Biophys. Acta.* **473**, 241–293.
11. Pegge, A. E. (1986). Recent advances in the biochemistry of polyamines in eukaryotes. *Biochem. J.* **234**, 249–262.
12. Wiberg, K. B. (1964). *Physical Organic Chemistry*, pp. 280–282. John Wiley, New York.
13. Liappis, N. (1972). Geschlechtsspezifische Unterschiede der freien Aminosauren im Serum von Erdwachsenen. *Z. Klin. Chem. Klin. Biochem.* **10**, 132–135.
14. Pegg, A. E. & McCann, P. P. (1982). Polyamine metabolism and function. *Am. J. Physiol.* **243**, C212–C221.
15. Tabor, C. W. & Tabor, H. (1985). Polyamines in microorganisms. *Microbiol. Rev.* **49**, 81–99.
16. Slocum, R. D., Kaur-Sawhney, R. & Galston, A. (1984). The physiology and biochemistry of polyamines in plants. *Arch. Biochem. Biophys.* **235**, 283–303.
17. Smith, T. A. (1985). Polyamines. *Ann. Rev. Plant Physiol.* **36**, 117–143.
18. International Union of Biochemistry (1984). *Enzyme Nomenclature.* Academic Press, New York and London.
19. Srivenugopal, K. S. & Adiga, P. R. (1981). Enzymic conversion of agmatine to putrescine in *Lathyrus sativus. J. Biol. Chem.* **256**, 9532–9541.
20. Pegg, A. E., Shuttleworth, K. & Hibasami, H. (1981). Specificity of mammalian spermidine synthase and spermine synthase. *Biochem. J.* **197**, 315–320.
21. Williams-Ashman, H. G., Seidenfeld, J. & Galletti, P. (1982). Trends in the biochemical pharmacology of 5'-deoxy-5'-methylthioadenosine. *Biochem. Pharmacol.* **31**, 277–288.
22. Schlenk, F. (1983). Methylthioadenosine. *Adv. Enzymol. Relat. Areas Mol. Biol.* **54**, 195–266.
23. Tait, G. H. (1985). Bacterial polyamines, structures and biosynthesis. *Biochem. Soc. Trans.* **13**, 316–318.

24. Srivenugopal, K. S. & Adiga, P. R. (1980). Coexistence of two pathways of spermidine biosynthesis in *Lathyrus sativus* seedlings. *FEBS Lett.* **112**, 260–264.
25. Seiler, N., Bolkenius, F. N. & Rennert, D. M. (1981). Interconversion, catabolism and elimination of the polyamines. *Med. Biol.* **59**, 334–346.
26. Pegg, A. E., Seely, J. E., Pösö, H., Della Ragione, F. & Zagon, I. S. (1982). Polyamine biosynthesis and interconversion in rodent tissues. *Fed. Proc.* **41**, 3065–3072.
27. Seely, J. E., Pösö, H. & Pegg, A. E. (1982). Purification of ornithine decarboxylase from kidneys of androgen-treated mice. *Biochemistry* **21**, 3394–3399.
28. Kameji, T., Murakami, Y., Fujita, K. & Hayashi, S. (1982). Purification and some properties of ornithine decarboxylase from rat liver. *Biochim. Biophys. Acta* **717**, 111–117.
29. Persson, L. (1982). Antibodies to ornithine decarboxylase. Immunochemical cross reactivity. *Acta Chem. Scand. Ser. B* **36**, 685–688.
30. Isomaa, V. V., Pajunen, A. E. I., Bardin, C. W. & Janne, D. A. (1983). Ornithine decarboxylase in mouse kidneys: purification, characterisation, and radioimmunological detection of the enzyme. *J. Biol. Chem.* **258**, 6735–6740.
31. Seely, J. E. G. & Pegg, A. E. (1983). Changes in mouse kidney ornithine decarboxylase activity are brought about by changes in the amount of enzyme protein as measured by radioimmunoassay. *J. Biol. Chem.* **258**, 2496–2500.
32. Metcalf, B. W., Bey, P., Danzin, C., Jung, M. J., Casara, P. & Vevert, J. P. (1978). Catalytic irreversible inhibition of mammalian ornithine decarboxylase (EC 4.1.1.17) by substrate and product analogues. *J. Am. Chem. Soc.* **100**, 2551–2553.
33. McConlogue, L., Gupta, M., Wu, L. & Coffino, P. (1984). Molecular cloning and expression of the mouse ornithine decarboxylase gene. *Proc. Natl. Acad. Sci. USA* **81**, 540–544.
34. Kontula, K. K., Torkkeli, T. J., Bardin, C. W. & Janne, D. A. (1984). Androgen induction of ornithine decarboxylase mRNA in mouse kidney as studied by complementary DNA. *Proc. Natl. Acad. Sci. USA* **81**, 731–735.
35. Kahana, C. & Nathans, D. (1984). Isolation of cloned cDNA encoding mammalian ornithine decarboxylase. *Proc. Natl. Acad. Sci. USA* **81**, 3645–3649.
36. Gupta, M. & Coffino, P. (1985). Mouse ornithine decarboxylase: complete amino acid sequence deduced from cDNA. *J. Biol. Chem.* **260**, 2941–2944.
37. Kahana, C. & Nathans, D. (1985). Nucleotide sequence of murine ornithine decarboxylase mRNA. *Proc. Natl. Acad. Sci. USA* **82**, 1673–1677.
38. Canellakis, E. S., Kyriakidis, D. A., Rinehart, C. A., Huang, S.-C., Panagiotidis, C. & Fong, W.-F. (1985). Regulation of polyamine biosynthesis by antizyme and some recent developments relating the induction of polyamine biosynthesis to cell growth. *Bioscience Rep.* **5**, 189–204.
39. Tabor, C. W. & Tabor, H. (1984). Methionine adenosyltransferase (*S*-adenosylmethionine synthase) and *S*-adenosylmethionine decarboxylase. *Adv. Enzymol. Relat. Areas Mol. Biol.* **56**, 251–282.
40. Morgan, D. M. L. (1983). Polyamine oxidation and human pregnancy. *Adv. Polyamine Res.* **4**, 155–167.
41. Pösö, H. & Pegg, A. E. (1982). Comparison of *S*-adenosylmethionine decarboxylase from rat liver and muscle. *Biochemistry* **21**, 3116–3122.
42. Libby, P. R. (1978). Calf liver nuclear *N*-acetyltransferases: purification and properties of two enzymes with both spermidine acetyltransferase and histone acetyltransferase activities. *J. Biol. Chem.* **253**, 233–237.

43. Erwin, B. G., Persson, L. & Pegg, A. E. (1984). Differential inhibition of histone and polyamine acetylases by multi-substrate analogues. *Biochemistry* **23**, 4250–4255.
44. Della Ragione, F. & Pegg, A. E. (1982). Purification and characterization of spermidine/spermine $N^1$-acetyltransferase from rat liver. *Biochemistry* **21**, 6152–6158.
45. Della Ragione, F. & Pegg, A. E. (1983). Studies of the specificity and kinetics of rat liver spermidine/spermine $N^1$-acetyltransferase. *Biochem. J.* **213**, 701–706.
46. Russell, D. M. (1985). Ornithine decarboxylase: a key regulatory enzyme in normal and neoplastic growth. *Drug Metabolism Rev.* **15**, 1–88.
47. Matsui-Yuasa, I., Otani, S., Morisawa, S., Takigawa, M., Enomoto, M. & Suzuki, F. (1985). Induction of spermidine/spermine $N^1$-acetyltransferase by parathyroid hormone in rabbit costal chondrocytes in culture. *J. Biochem. Tokyo* **97**, 387–390.
48. Matsui-Yuasa, I., Otani, S. & Morisawa, S. (1985). Calcium ionophore A23187 induction of spermidine/spermine $N^1$-acetyltransferase activity in bovine lymphocytes. *FEBS Lett.* **188**, 375–378.
49. Erwin, B. G. & Pegg, A. E. (1986). Regulation of spermidine/spermine $N^1$-acetyltransferase in L6 cells by polyamines and related compounds. *Biochem. J.* **238**, 581–587.
50. Bachrach, U. (1970). Oxidised polyamines. *Ann. N.Y. Acad. Sci.* **171**, 939–956.
51. Bachrach, U. (1981). Oxidation of polyamines and diamines. In *Polyamines in Biology and Medicine* (Morris, D. R. & Marton, L. J., eds), pp. 151–168. Marcel Dekker, New York and Basel.
52. Morgan, D. M. L. (1980). Polyamine oxidases. In *Polyamines in Biomedical Research* (Gaugas, J. M., ed.), pp. 285–302. John Wiley, Chichester.
53. Morgan, D. M. L. (1985). Polyamine oxidases. *Biochem. Soc. Trans.* **13**, 322–326.
54. Morgan, D. M. L. (1988). Polyamine oxidases and oxidised polyamines. In *The Physiology of the Polyamines* (Bachrach, U. & Heimer, Y., eds). CRC Press, Boca Raton (in press).
55. Yamada, H. & Yasunobu, K. T. (1962). Monoamine oxidase. I. Purification, crystallisation, and properties of plasma monoamine oxidase. *J. Biol. Chem.* **237**, 1511–1516.
56. Yasunobu, K. T., Ishizaki, H. & Minamura, N. (1976). The molecular, mechanistic and immunological properties of amine oxidases. *Mol. Cell. Biochem.* **13**, 3–29.
57. Turini, P., Sabatini, S., Betani, O., Chimenti, F., Casanova, C., Riccio, P. L. & Mondovi, B. (1982). Purification of bovine plasma amine oxidase. *Analyt. Biochem.* **125**, 294–298.
58. Summers, M. C., Markovic, R. & Klinman, J. P. (1979). Stereochemistry and kinetic isotope effects in the bovine plasma amine oxidase catalysed oxidation of dopamine. *Biochemistry* **18**, 1969–1979.
59. Tabor, C. W., Tabor, H. & Bachrach, U. (1964). Identification of the amino-aldehydes produced by the oxidation of spermine and spermidine with purified plasma amine oxidase. *J. Biol. Chem.* **239**, 2194–2203.
60. Achee, F., Chervenka, C. H., Smith, R. A. & Yasunobu, K. T. (1968). Amine oxidase. XII. The association and dissociation, and number of subunits, of beef plasma amine oxidase. *Biochemistry* **7**, 4329–4335.

61. Yamada, H., Gee, P., Ebata, M. & Yasunobu, K. T. (1964). Monoamine oxidase. VI. Physicochemical properties of plasma monoamine oxidase. *Biochem. Biophys. Acta* **81**, 165–171.
62. Nakano, G., Harada, M. & Nagatsu, T. (1974). Purification and properties of an amine oxidase in bovine dental pulp and its comparison with serum amine oxidase. *Biochim. Biophys. Acta* **341**, 366–377.
63. Gahl, W. A. & Pitot, H. C. (1982). Polyamine degradation in foetal and adult bovine serum. *Biochem. J.* **202**, 603–611.
64. Mondovi, B. B., Sabatini, S. & Befani, O. (1984). Recent studies on the active site of beef plasma amine oxidase. In *Advances in Polyamines in Biomedical Science* (Caldarera, C. M. & Bachrach, U., eds), pp. 75–79. CLUEB, Bologna.
65. Yamada, H. & Yasunobu, K. T. (1963). Monoamine oxidase. IV. Nature of the second prosthetic group of plasma monoamine oxidase. *J. Biol. Chem.* **238**, 2669–2675.
66. Suva, R. H. & Abeles, R. H. (1978). Studies on the mechanism of action of plasma amine oxidase. *Biochemistry* **17**, 3538–3545.
67. Lobenstein-Verbeek, C. L., Jongjan, J. A., Frank, J. & Duine, J. A. (1984). Bovine serum amine oxidase: a mammalian enzyme having covalently bound PQQ as prosthetic group. *FEBS Lett.* **170**, 305–309.
68. Duine, J. A. & Franke, J. (1981). Quinoproteins: a novel class of dehydrogenases. *Trends Biochem. Sci.* **6**, 278–280.
69. Watanabe, K. & Yasunobu, K. T. (1970). Carbohydrate content of bovine plasma amine oxidase and isolation of a carbohydrate containing fragment attached to asparagine. *J. Biol. Chem.* **245**, 4612–4617.
70. Ishsizaki, H. & Yasunobu, K. T. (1980). Bovine plasma amine oxidase interactions with concanavalin A in solution and with concanavalin A-Sepharose. *Biochim. Biophys. Acta* **611**, 27–34.
71. Israel, M., Rosenfield, J. S. & Modest, E. J. (1964). Analogs of spermine and spermidine. I. Synthesis of polymethylene-polyamines by reduction of cyanoethylated α,ω-alkylenediamines. *J. Med. Chem.* **7**, 710–716.
72. Tabor, C. W., Tabor, H. & Rosenthal, S. M. (1954). Purification of amine oxidase from beef plasma. *J. Biol. Chem.* **208**, 645–661.
73. Wang, T.-M., Achee, F. M. & Yasunobu, K. T. (1968). Amine oxidase. XI. Beef plasma amine oxidase, a sulphhydryl protein. *Arch. Biochem. Biophys.* **128**, 106–112.
74. Yamada, H. & Yasanobu, K. T. (1962). Monoamine oxidase II. Copper, one of the prosthetic groups of plasma monoamine oxidase. *J. Biol. Chem.* **237**, 3077–3082.
75. Levine, C. I. & Gross, J. (1959). Alterations in state of molecular aggregation of collagen induced in chick embryos by β-aminopropionitrile (lathyrus factor). *J. Exp. Med.* **110**, 771–790.
76. Page, R. C. & Benditt, E. P. (1967). Interaction of the lathyrogen beta-aminopropionitrile (BAPN) with a copper-containing amine oxidase. *Proc. Soc. Exp. Biol. Med.* **124**, 454–459.
77. Crabbe, M. J. C. (1979). Histaminases in human placenta and seminal fluid and their possible similarities to lysyl oxidase. *Agents Actions* **9**, 41–42.
78. Blaschko, H. (1962). The amine oxidases of mammalian blood plasma. *Adv. Comp. Physiol. Biochem.* **1**, 67–116.
79. Seiler, N., Knodgen, B., Gittos, M. W., Chan, W. Y., Griesmann, G. & Rennert, O. M. (1981). On the formation of amino acids deriving from spermidine and spermine. *Biochem. J.* **200**, 123–132.

80. Seiler, N., Knodgen, B., Bink, G., Sarhan, S. & Bolkenius, F. (1983). Diamine oxidase and polyamine catabolism. *Adv. Polyamine Res.* **4**, 135–154.
81. Seiler, N. & Knodgen, B. (1983). *N*-(3-aminopropyl)-pyrrolidin-2-one: a physiological excretory product deriving from spermidine. *Int. J. Biochem.* **15**, 907–915.
82. Asatoor, A. M. (1979). Isolation and characterisation of a new urinary metabolite derived from spermidine. *Biochim. Biophys. Acta* **586**, 55–62.
83. Noto, T., Tanaka, T. & Nakajima, T. (1978). Urinary metabolites of polyamines in rats. *J. Biochem. Tokyo* **83**, 543–552.
84. Nakajima, T., Matsuoka, Y. & Akazawa, S. (1970). Putreanine excretion in human urine. *Biochim. Biophys. Acta* **222**, 405–408.
85. Van Den Berg, G. A., Elzinga, H., Nagel, G. T., Kingma, A. W. & Muskiet, F. A. J. (1984). Catabolism of polyamines in the rat. Polyamines and their non-alpha-amino-acid metabolites. *Biochim. Biophys. Acta* **802**, 175–187.
86. Höltta, E., Pulkkinen, P., Elfving, K. & Janne, J. (1975). Oxidation of polyamines by diamine oxidase from human seminal plasma. *Biochem. J.* **145**, 373–378.
87. Höltta, E. (1977). Oxidation of spermidine and spermine in rat liver: purification and properties of polyamine oxidase. *Biochemistry* **16**, 91–100.
88. Höltta, E. (1983). Polyamine oxidase (rat liver). *Meth. Enzymol.* **94**, 306–311.
89. Beard, M. E., Baker, R., Conomos, P., Pugatch, D. R. & Holtzman, E. (1985). Oxidation of oxalate and polyamines by rat peroxisomes. *J. Histochem. Cytochem.* **33**, 460–464.
90. Bolkenius, F. N. & Seiler, N. (1981). Acetylderivatives as intermediates in polyamine catabolism. *Int. J. Biochem.* **13**, 287–292.
91. Seiler, N., Bolkenius, F. N., Knodgen, B. & Mamont, P. (1980). Polyamine oxidase in rat tissues. *Biochim. Biophys. Acta* **615**, 480–488.
92. Illei, G. & Morgan, D. M. L. (1979). The distribution of polyamine oxidase activity in the fetomaternal compartments. *Br. J. Obstet. Gynaecol.* **86**, 873–877.
93. Illei, G. & Morgan, D. M. L. (1980). Polyamine oxidase activity in amniotic fluid and fetal membranes. *Br. J. Obstet. Gynaecol.* **87**, 413–415.
94. Suzuki, O., Matsumoto, T. & Katsumata, Y. (1984). Determination of polyamine oxidase activities in human tissues. *Experientia* **40**, 838–839.
95. Tabor, H. (1954). Metabolic studies on histidine, histamine, and related imidazoles. *Pharmacol. Rev.* **6**, 299–343.
96. Buffoni, F. (1966). Histaminase and related amine oxidases. *Pharmacol. Rev.* **18**, 1163–1199.
97. Ahlmark, A. (1944). Studies on the histaminolytic power of plasma with special reference to pregnancy. *Acta Physiol. Scand.* **9** (Suppl. 28), 1–107.
98. Southren, A. L., Kobayashi, Y., Carmody, N. C. & Weingold, A. B. (1966). Serial measurements of plasma diamine oxidase (DAO) during normal human pregnancy by an improved method. *Am. J. Obstet. Gynaecol.* **95**, 615–620.
99. Gahl, W. A., Vale, A. M. & Pitot, H. C. (1982). Spermidine oxidase in human pregnancy serum. Probable identity with diamine oxidase. *Biochem. J.* **201**, 161–164.
100. Illei, G. & Morgan, D. M. L. (1979). Polyamine oxidase activity in human pregnancy serum. *Br. J. Obstet. Gynaecol.* **86**, 878–881.
101. Morgan, D. M. L., Illei, G. & Royston, J. P. (1983). Serum polyamine oxidase activity in normal pregnancy. *Br. J. Obstet. Gynaecol.* **90**, 1194–1196.

102. Lin, C.-W., Chapman, C. M., DeLellis, R. A. & Kinley, S. (1978). Immunofluorescent staining of histamines (diamine oxidase) in human placenta. *J. Histochem. Cytochem.* **26**, 1021–1025.
103. Weisburger, W. R., Mendelsohn, G., Eggleston, J. C. & Baylin, S. B. (1978). Immunohistochemical localisation of histaminase (diamine oxidase) in decidual cells of human placenta. *Lab. Invest.* **38**, 703–706.
104. Smith, J. K. (1967). The purification and properties of placental histaminase. *Biochem. J.* **103**, 110–119.
105. Paolucci, F., Cronenberger, L., Plan, R. & Pachsko, H. (1971). Purification and properties of the diamine: oxygen oxydo-reductase of human placenta. *Biochemie* **53**, 735–749.
106. Bardsley, W. G., Crabbe, M. J. C. & Scot, I. V. (1974). The amine oxidase of human placenta and pregnancy plasma. *Biochem. J.* **139**, 169–181.
107. Crabbe, M. J. C. & Bardsley, W. G. (1974). The inhibition of placental diamine oxidase by substrate analogs. *Biochem. J.* **139**, 183–189.
108. Crabbe, M. J. C. & Bardsley, W. G. (1974). Monoamine oxidase inhibitors and other drugs as inhibitors of diamine oxidase from human placenta and pig kidney. *Biochem. Pharmacol.* **23**, 2983–2990.
109. Crabbe, M. J. C., Childs, R. E. & Bardsley, W. G. (1975). Time dependent inhibition of diamine oxidase by carbonyl group reagents and urea. *Eur. J. Biochem.* **60**, 325–333.
110. Crabbe, M. J. C., Waight, R. D., Bardsley, W. G., Barker, R. W., Kelley, I. D. & Knowles, P. F. (1976). Human placental diamine oxidase. Improved purification and characterization of a copper- and manganese-containing amine oxidase with novel substrate specificity. *Biochem. J.* **155**, 679–687.
111. Morgan, D. M. L. (1985). Human pregnancy-associated polyamine oxidase: partial purification and properties. *Biochem. Soc. Trans.* **13**, 351–352.
112. Morgan, D. M. L. & Toothill, V. J. (1985). Polyamine oxidase activity in human and bovine milk. *Biochem. Soc. Trans.* **13**, 352–353.
113. Crabbe, M. J. C. (1981). Summary of an international workshop on diamine oxidase held at the Department of Biogenic Amines, Polish Academy of Sciences, Lodz, Poland, 7–8 June, 1980. *Agents Actions* **11**, 3–8.
114. Russell, D. H. (1983). Clinical relevance of polyamines. *CRC. Crit. Rev. Clin. Lab. Sci.* **18**, 261–311.
115. Hiramatsu, Y., Eguchi, K. & Sekiba, K. (1985). Alterations in polyamine levels in amniotic fluid, plasma and urine during normal pregnancy. *Acta Med. Okayama* **39**, 339–346.
116. Illei, G. & Morgan, D. M. L. (1982). Serum polyamine oxidase activity in spontaneous abortion. *Br. J. Obstet. Gynaecol.* **89**, 199–201.
117. Byrd, W. J., Jacobs, D. M. & Amos, M. S. (1977). Synthetic polyamines added to cultures containing bovine sera reversibly inhibit *in vitro* parameters of immunity. *Nature* **267**, 621–623.
118. Byrd, W. J., Jacobs, D. M. & Amos, M. S. (1978). Influence of synthetic polyamines on the *in vitro* responses of immunocompetent cells. *Adv. Polyamine Res.* **2**, 71–83.
119. Gaugas, J. M. & Curzen, P. (1978). Polyamine interaction with pregnancy serum in suppression of lymphocyte transformation. *Lancet* **i**, 18–20.
120. Morgan, D. M. L. & Illei, G. (1980). Polyamine–polyamine oxidase interaction: part of maternal protective mechanism against fetal rejection. *Br. Med. J.* **280**, 1295–1297.

121. Morgan, D. M. L. (1981). Polyamine oxidase in human pregnancy: a possible immunoregulatory agent. *Biochem. Soc. Trans.* **9**, 400–401.
122. Labib, R. S. & Tomasi, T. B. (1981). Enzymatic oxidation of polyamines. Relationship to immunosuppressive properties. *Eur. J. Immunol.* **11**, 266–269.
123. Khomenko, A. K., Dyachok, F. I., Gridina, N. Y., Burlaka, D. P., Negrey, G. Z. & Shylakhovenko, V. A. (1984). Influence of serum polyamine oxidase on normal and tumor cells. *Eksperimentainaya Onkologiya* **6**, 63–66.
124. Yamada, H., Isobe, K. & Tani, Y. (1980). Oxidation of polyamines by fungal enzymes. *Agric. Biol. Chem.* **44**, 2469–2476.
125. Kumagi, H. & Yamada, H. (1985). Bacterial and fungal amine oxidases. In *Structure and Functions of Amine Oxidases* (Mondovi, B., ed.), pp. 37–43. CRC Press, Boca Raton.
126. Isobe, K., Tani, Y. & Yamada, H. (1980). Crystallisation and characterization of polyamine oxidase from *Penicillium chrysogenum*. *Agric. Biol. Chem.* **44**, 2651–2658.
127. Isobe, K., Tani, Y. & Yamada, H. (1980). Crystallization and characterization of polyamine oxidase from *Aspergillus terreus*. *Agric. Biol. Chem.* **44**, 2479–2751.
128. Isobe, K., Tani, Y. & Yamada, H. (1980). Kinetic properties of fungal polyamine oxidases and their application to differential determination of spermine and spermidine. *Agric. Biol. Chem.* **44**, 2955–2960.
129. Haywood, G. W. & Large, P. J. (1984). Partial purification of a peroxisomal polyamine oxidase from *Candida boidinii* and its role in growth on spermidine as sole nitrogen source. *J. Gen. Microbiol.* **130**, 1123–1136.
130. Green, J., Haywood, G. W. & Large, P. J. (1983). Serological differences between the multiple amine oxidases of yeasts and comparison of the specificities of the purified enzymes from *Candida utilis* and *Pichia pastoris*. *Biochem. J.* **211**, 481–493.
131. Smith, T. A. (1985). The di- and polyamine oxidases of higher plants. *Biochem. Soc. Trans.* **13**, 319–322.
132. Rinaldi, A., Floris, G. & Giartosio, A. (1985). Plant amine oxidases. In *Structure and Functions of Amine Oxidases* (Mondovi, B., ed.), pp. 51–62. CRC Press, Boca Raton.
133. Croker, S. J., Loeffler, R. S. T. & Smith, T. A. (1983). 1,5-diazabicyclo[4.3.0.]nonane, the oxidation product of spermine. *Tetrahedron Lett.* **24**, 1559–1560.
134. Brandange, S. & Eriksson, L.-H. (1984). Ring-chain tautomerism of an *N,N*-acetal formed by enzymatic oxidation of spermine or spermidine. *Acta Chem. Scand., Ser. B* **38**, 526–528.
135. Suzuki, Y. & Yanagisawa, H. (1980). Purification and properties of maize polyamine oxidase—a flavoprotein. *Plant Cell Physiol.* **21**, 1085–1094.
136. Cohen, S. S. (1978). The functions of the polyamines. *Adv. Polyamine Res.* **1**, 1–10.
137. Cohen, S. S. (1978). What do the polyamines do? *Nature* (*London*) **274**, 209–210.
138. Abraham, A. K. & Pihl, A. (1981). Role of polyamines in macromolecular synthesis. *Trends Biochem. Sci.* **6**, 106–107.
139. Marton, L. J. & Feuerstein, B. G. (1986). Polyamine–DNA interactions: possible site of new cancer chemotherapeutic intervention. *Pharmaceut. Res.* **3**, 311–317.

140. Heby, O. (1981). Role of polyamines in the control of cell proliferation and differentiation. *Differentiation* **19**, 1–20.
141. Rossow, P. W., Riddle, V. G. H. & Pardee, A. B. (1979). Synthesis of a labile, serum-dependent protein in early G controls animal cell growth. *Proc. Natl. Acad. Sci. USA* **76**, 4446–4450.
142. Kano, K. & Oka, T. (1976). Polyamine transport and metabolism in mouse mammary gland: general properties and hormonal regulation. *J. Biol. Chem.* **251**, 2795–2800.
143. Gordonsmith, R. H., Brooke-Taylor, S., Smith, L. L. & Cohen, G. M. (1983). Structural requirements of compound to inhibit pulmonary diamine accumulation. *Biochem. Pharmacol.* **32**, 3701–3709.
144. Alhonen-Hongisto, L., Seppanen, P. & Janne, J. (1980). Intracellular putrescine and spermidine deprivation induces increased uptake of the natural polyamines and methyglyoxal bis(guanylhydrazone). *Biochem. J.* **192**, 941–945.
145. Smith, L. L. (1985). The identification and characterisation of a polyamine-accumulation system in the lung. *Biochem. Soc. Trans.* **13**, 332–334.
146. Folk, J. E. (1980). Transglutaminases. *Ann. Rev. Biochem.* **49**, 517–531.
147. Scalabrino, G. & Ferioli, M. E. (1981). Polyamines in mammalian tumours. Part I. *Adv. Cancer Res.* **35**, 151–268.
148. Aigner-Held, R. & Daves, G. D. (1980). Polyamine metabolites and conjugates in man and higher animals: a review of the literature. *Physiol. Chem. Phys.* **12**, 389–400.
149. Seiler, N., Bolkenius, F. N. & Knodgen, B. (1980). Acetylation of spermidine in polyamine catabolism. *Biochim. Biophys. Acta* **633**, 181–190.
150. Morgan, D. M. L. (1987). Oxidised polyamine and the growth of human vascular endothelial cells: prevention of cytotoxic effects by selective acetylation. *Biochem. J.* **242**, 347–352.
151. Morgan, D. M. L., Bachrach, U., Assaraf, Y., Harari, E. & Golenser, J. (1986). The effect of purified aminoaldehydes produced by polyamine oxidation on the *in vitro* development of *Plasmodium falciparum* in normal and glucose-6-phosphate-dehydrogenase deficient erythrocytes. *Biochem. J.* **236**, 97–101.
152. Allen, J. C. & Smith, C. J. (1979). Chalones: a reappraisal. *Biochem. Soc. Trans.* **7**, 584–592.
153. Patt, L. M. & Houck, J. C. (1980). The incredible shrinking chalone. *FEBS Lett.* **120**, 163–170.
154. Allen, J. C., Smith, C. J., Curry, M. C. & Gaugas, J. M. (1977). Identification of a thymic inhibitor ('chalone') of lymphocyte transformation as a spermine complex. *Nature* **267**, 623–625.
155. Rytomaa, T. (1976). The chalone concept. *Int. Rev. Pathol.* **16**, 155–206.
156. Quash, G., Keolouankhot, T., Gazzolo, L., Ripoll, H. & Saez, S. (1979). Diamine and polyamine oxidase activities in normal and transformed cells. *Biochem. J.* **177**, 275–282.
157. Kingsnorth, A. N. & Wallace, H. M. (1985). Elevation of monoacetylated polyamines in human breast cancers. *Eur. J. Cancer Clin. Oncol.* **21**, 1057–1062.
158. Mach, M., Kersten, H. & Kersten, W. (1982). Regulation of tRNA methyltransferase activities by spermidine and putrescine. Inhibition of polyamine synthesis and tRNA methylation by $\alpha$-methylornithine and 1,3-diaminopropan-2-ol in *Dictyostelium discoideum*. *Biochem. J.* **202**, 153–162.

159. Morgan, D. M. L., Ferluga, J. & Allison, A. C. (1980). Polyamine oxidase and macrophage function. In *Polyamines in Biomedical Research* (Gaugas, J. M., ed.), pp. 303–308. John Wiley, Chichester and New York.
160. Smith, C. J., Maschler, R., Maurer, H. P. & Allen, J. C. (1983). Inhibition of cells in culture by polyamines does not depend on the presence of ruminant serum. *Cell Tissue Kinet.* **16**, 269–276.
161. Frolik, C. A., Roller, P. P., Cone, J. L., Dart, L. L., Smith, D. M. & Sporn, M. B. (1984). Inhibition of transforming growth factor-induced cell growth in soft agar by oxidised polyamines. *Arch. Biochem. Biophys.* **230**, 93–102.
162. Mezl, V. A., Fournier, L. A. & Garber, P. M. (1986). $N^1$-amino-acetylation abolishes the inhibitory effect of spermine and spermidine in the reticulolcyte lysate translation system. *Int. J. Biochem.* **18**, 705–711.
163. Pegg, A. E. & Erwin, B. G. (1985). Induction of spermidine/spermine $N^1$-acetyltransferase in rat tissues by polyamines. *Biochem. J.* **231**, 285–289.
164. Porter, C. W., Bergeron, R. J. & Stolowich, N. J. (1982). Biological properties of $N^4$-spermidine derivatives and their potential in anticancer chemotherapy. *Cancer Res.* **42**, 4072–4078.
165. Porter, C. W., Cavanaugh, P. F., Stolowich, N., Ganis, B., Kelly, E. & Bergeron, R. J. (1985). Biological properties of $N^4$- and $N^1,N^8$-spermidine derivatives in cultured L1210 leukemia cells. *Cancer Res.* **45**, 2050–2057.
166. Seiler, N. (1981). Amide-bond-forming reactions of polyamines. In *Polyamines in Biology and Medicine* (Morris, D. R. & Marton, L. J., eds), pp. 127–149. Marcel Dekker, Basel and New York.
167. Wallace, H. M. & Keir, H. M. (1986). Factors affecting polyamine excretion from mammalian cells in culture. *FEBS Lett.* **194**, 60–63.
168. Seiler, N. (1985). Acetylpolyamines as substrates of amine oxidases. In *Structure and Functions of Amine Oxidases* (Mondovi, B., ed.), pp. 21–35. CRC Press, Boca Raton.
169. Bey, P., Bolkenius, F. N., Seiler, N. & Casara, P. (1985). *N*-2,3-Butadienyl-1,4-butanediamine derivatives: potent irreversible inactivators of mammalian polyamine oxidase. *J. Med. Chem.* **28**, 1–2.
170. Bolkenius, F. N., Bey, P. & Seiler, N. (1985). Specific inhibition of polyamine oxidase *in vivo* is a method for the elucidation of its physiological role. *Biochem. Biophys. Acta* **839**, 69–76.
171. Seiler, N., Bolkenius, F. N. & Knodgen, B. (1985). The influence of catabolic reactions on polyamine excretion. *Biochem. J.* **225**, 219–226.
172. Seiler, N. & Bolkenius, F. N. (1985). Polyamine reutilization and turnover in brain. *Neurochem. Res.* **10**, 529–544.
173. Scalabrino, G. & Ferioli, M. E. (1982). Polyamines in mammalian tumors. Part II. *Adv. Cancer. Res.* **36**, 1–102.
174. Oredsson, S. M. & Marton, J. (1982). Polyamines: the elusive cancer markers. *Clinics Lab. Med.* **2**, 507–518.
175. Seiler, N. (1986). Polyamines. *J. Chromatogr.* **379**, 157–176.
176. Womble, J. R., Larson, D. F., Copeland, J. G. & Russell, D. H. (1984). Urinary polyamine levels are markers of altered T-lymphocyte proliferation/loss and rejection in heart transplant patients. *Transplant Proc.* **16**, 1573–1575.
177. Porter, C. W. & Sufrin, J. R. (1986). Interference with polyamine biosynthesis and/or function by analogs of polyamines or methionine as a potential anticancer chemotherapeutic strategy. *Anticancer Res.* **6**, 525–542.
178. Morgan, D. M. L., Christensen, J. R. & Allison, A. C. (1981). Polyamine

oxidase and the killing of intracellular parasites. *Biochem. Soc. Trans.* **9**, 563–564.

179. Morgan, D. M. L. & Christensen, J. R. (1983). Polyamine oxidation and the killing of intracellular parasites. *Adv. Polyamine Res.* **4**, 169–174.

180. Ferrante, A., Allison, A. C. & Hirumi, H. (1982). Polyamine oxidase-mediated killing of African trypanosomes. *Parasite Immunol.* **4**, 349–354.

181. Ferrante, A., Rzepczyk, A. & Allison, A. C. (1983). Polyamine oxidase mediates intra-erythrocytic death of *Plasmodium falciparum. Trans. Roy. Soc. Trop. Med. Hyg.* **77**, 789–791.

182. Ferrante, A., Rzepczyk, C. M. & Saul, A. J. (1984). Polyamine oxidase-mediated trypanosome killing: the role of hydrogen peroxide and aldehydes. *J. Immunol.* **133**, 2157–2162.

183. Rzepczyk, C. M., Saul, A. J. & Ferrante, A. (1984). Polyamine oxidase-mediated intraerythrocytic killing of *Plasmodium falciparum*: evidence against the role of reactive oxygen metabolites. *Infect. Immunol.* **43**, 238–244.

184. Egan, J. E., Haynes, J. D., Brown, N. D. & Eismann, C. S. (1986). Human polyamine oxidase inhibits *P. falciparum. Am. J. Trop. Med. Hyg.* **35**, 890–897.

185. Wang, C. C. (1984). Parasite enzymes as potential targets for antiparasitic chemotherapy. *J. Med. Chem.* **27**, 1–9.

186. Sjoerdsma, A. & Schechter, P. J. (1984). Chemotherapeutic implications of polyamine biosynthesis inhibition. *Clin. Pharmacol. Therapeutics* **35**, 287–300.

187. McCann, P. P., Bacchi, C. J., Clarkson, A. B., Bey, P., Sjoerdsma, A., Schechter, P. J., Walzer, P. D. & Barlow, J. L. R. (1986). Inhibition of polyamine biosynthesis by difluoromethylornithine in African trypanosomes and *Pneumocystis carinii* as a basis of chemotherapy: biochemical and clinial aspects. *Am. J. Trop. Med. Hyg.* **35**, 1153–1156.

188. Ferrante, A., Ljungstrom, I., Rzepczyk, C. & Morgan, D. M. L. (1986). Differences in sensitivity of *Schistosoma mansoni* schistosomula, *Dirofilaria immitis* microfilariae, and *Nematospiroides dubius* third stage larvae to damage by the polyamine oxidase-polyamine system. *Infect. Immun.* **53**, 606–610.

189. Mamont, P. S., Bohlen, P., McCann, P. P., Bey, P., Schuber, F. & Tardif, C. (1976). α-Methyl ornithine, a potent competitive inhibitor of ornithine decarboxylase, blocks proliferation of rat hepatoma cells in culture. *Proc. Natl. Acad. Sci. USA* **73**, 1626–1630.

190. Smith, T. A. (1985). Introductory chapter—polyamines in plants. In *Polyamines in Plants* (Galston, A. W. & Smith, T. A., eds), pp. vii–xxi. Martinus Nijhoff, Dordrecht.

191. Tabor, H., Tabor, C. W. & De Meis, L. (1971). Chemical synthesis of $N$-acetyl-1,4-diaminobutane, $N^1$-acetylspermidine, and $N^8$-acetylspermidine. *Meth. Enzymol.* **17B**, 829–833.

192. Onasch, F., Aikens, D., Bunce, S., Schwarts, H., Nairn, D. & Hurwitz, C. (1984). The interactions between nucleic acids and polyamines. III. Microscopic protonation constants of spermidine. *Biophys. Chem.* **19**, 245–253.

193. Aikens, D. A., Bunce, S., Onasch, F., Parker, R., Hurwitz, C. & Clemans, S. (1983). The interactions between nucleic acids and polyamines. II. Protonation constants and $^{13}$C-NMR chemical shift assignments of spermidine, spermine, and homologs. *Biophys. Chem.* **17**, 67–74.

194. Adachi, O., Yamada, H. & Ogata, K. (1966). Purification and properties of putrescine oxidase of *Micrococcus rubens. Agric. Biol. Chem.* **30**, 1202–1210.

# Structure and Regulation of Glutathione S-Transferase Genes

## CECIL B. PICKETT

*Department of Molecular Pharmacology and Biochemistry, Merck Sharp and Dohme Research Laboratories, Rahway, New Jersey 07065, USA*

## I. Introduction

The glutathione S-transferases comprise a family of proteins that catalyse the conjugation of glutathione to various electrophilic ligands. In addition to the conjugation activity, the transferases bind a number of hydrophobic

ESSAYS IN BIOCHEMISTRY Vol. 23
ISBN 0 12 158123-3

compounds such as heme, bilirubin, polycyclic aromatic hydrocarbons and dexamethasone.[1-8] The glutathione *S*-transferases are heterodimers or homodimers comprising at least seven subunits (Ya, $Yb_1$, $Yb_2$, Yc, Yp, Yk and Yn).[9-13] An alternative nomenclature[14] has been proposed for the various glutathione *S*-transferases (Ya = subunit 1, Yc = subunit 2, $Yb_1$ = subunit 3, $Yb_2$ = subunit 4, Yn = subunit 6, Yp = subunit 7, Yk = subunit 8. However, in this review, the nomenclature originally suggested by Bass *et al.*[9] will be used. Some of the glutathione *S*-transferases are induced by xenobiotics (e.g. phenobarbital and 3-methylcholanthrene) as well as being elevated in persistent hepatocyte nodules induced by chemical carcinogens. Furthermore, numerous studies have suggested that the glutathione *S*-transferases are expressed in a tissue-specific manner.

This review focusses on how molecular biological approaches have begun to unravel the complexity of glutathione *S*-transferase subunits and their expression. Since much of this work has been done in the rat, many of the studies discussed here will focus on this experimental animal. For a discussion on the purification and enzymology of various purified glutathione *S*-transferases, the reader is referred to other recent reviews or book chapters.[2,3,15,16]

## II. Rat Liver Glutathione *S*-Transferase cDNAs

### A. NUCLEOTIDE SEQUENCE ANALYSIS OF THE RAT LIVER GLUTATHIONE *S*-TRANSFERASE Ya AND Yc cDNAs AND AMINO ACID SEQUENCES OF THE Ya AND Yc SUBUNITS

Recently, cDNA clones complementary to rat liver glutathione *S*-transferase Ya and Yc mRNAs have been constructed.[17-24] DNA sequence analysis of these clones have revealed that the Ya and Yc mRNAs are 66% identical. In the protein coding region, the nucleotide sequence identity is 75%. However, both the 5' and 3' untranslated regions of the two mRNAs are very divergent. The divergence in the 5' and 3' flanking regions of the Ya and Yc mRNAs as well as the random nucleotide differences throughout the Ya and Yc cDNA sequences indicate that these two mRNAs are products of separate genes. The nucleotide sequences for the Ya and Yc clones determined by our laboratory is presented in Fig. 1.

The amino acid sequences of the Ya and Yc subunits have an overall identity of 68%. The Ya subunit is comprised of 222 amino acids with a molecular weight of 25 547; whereas the Yc subunit is comprised of 221 amino acids with a molecular weight of 25 322. The amino acid sequences of the Ya and Yc subunits are presented in Fig. 2.

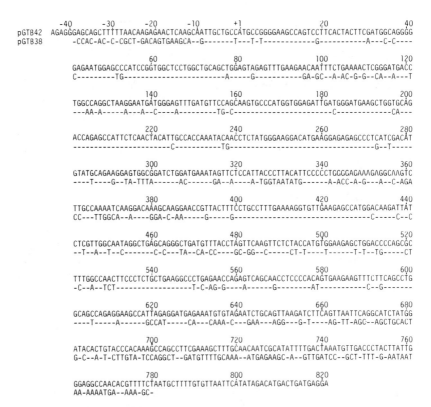

**Fig. 1.** DNA sequence analysis of pGTB42 (Yc clone) and pGTB38 (Ya clone). Dashed lines in the pGTB38 sequence represent nucleotides that are identical to those in the pGTB42 sequence.

It is also clear from cDNA sequence data that more than one Ya gene is expressed in rat liver. Three full-length or nearly full-length Ya cDNA clones have been isolated and characterized to date. Two clones, pGTB38[19] and pGTR261[18] differ by 15 nucleotides, which translate into eight amino acid differences. The 3' untranslated regions of the two cDNA clones are divergent, which support the idea that the two clones are derived from distinct mRNAs encoded by separate genes. A third Ya cDNA, pGTB45, is more similar to pGTR261 than pGTB38.[25] However, pGTB45 contains a Type-2 Alu repetitive element in the 3' untranslated region. The type-2 Alu repetitive element contains two overlapping polyadenylation signals downstream from the polyadenylation signal used in pGTR261. The functional significance of the repetitive element is unknown at the present time but may play a role in the expression or stability of this Ya mRNA.

```
                                     9                                    18
Yc MET Pro Gly Lys Pro Val Leu His Tyr Phe Asp Gly Arg Gly Arg Met Glu Pro
Ya MET Ser Gly Lys Pro Val Leu His Tyr Phe Asn Ala Arg Gly Arg Met Glu Cys

                                    27                                    36
   Ile Arg Trp Leu Leu Ala Ala Ala Gly Val Glu Phe Glu Glu Gln Phe Leu Lys
   Ile Arg Trp Leu Leu Ala Ala Ala Gly Val Glu Phe Glu Glu Lys Leu Ile Gln

                                    45                                    54
   Thr Arg Asp Asp Leu Ala Arg Leu Arg Asn Asp Gly Ser Leu Met Phe Gln Gln
   Ser Pro Glu Asp Leu Glu Lys Leu Lys Lys Asp Gly Asn Leu Met Phe Asp Gln

                                    63                                    72
   Val Pro Met Val Glu Ile Asp Gly Met Lys Leu Val Gln Thr Arg Ala Ile Leu
   Val Pro Met Val Glu Ile Asp Gly Met Lys Leu Ala Gln Thr Arg Ala Ile Lue

                                    81                                    90
   Asn Tyr Ile Ala Thr Lys Tyr Asn Leu Tyr Gly Lys Asp Met Lys Glu Arg Ala
   Asn Tyr Ile Ala Thr Lys Tyr Asp Leu Tyr Gly Lys Asp Met Lys Glu Arg Ala

                                    99                                   108
   Leu Ile Asp Met Tyr Ala Glu Gly Val Ala Asp Leu Asp Glu Ile Val Leu His
   Leu Ile Asp Met Tyr Ser Glu Gly Ile Leu Asp Leu Thr Glu Met Ile Ile Gln

                                   117                                   126
   Tyr Pro Tyr Ile Pro Pro Gly Glu Lys Glu Ala Ser Leu Ala Lys Ile Lys Asp
   Leu Val Ile Cys Pro Pro Asp Gln Arg Glu Ala Lys Thr Ala Leu Ala Lys Asp

                                   135                                   144
   Lys Ala Arg Asn Arg Tyr Phe Pro Ala Phe Glu Lys Val Leu Lys Ser His Gly
   Arg Thr Lys Asn Arg Tyr Leu Pro Ala Phe Glu Lys Val Leu Lys Ser His Gly

                                   153                                   162
   Gln Asp Tyr Leu Val Gly Asn Arg Leu Ser Arg Ala Asp Val Tyr Leu Val Gln
   Gln Asp Tyr Leu Val Gly Asn Arg Leu Thr Arg Val Asp Ile His Leu Leu Glu

                                   171                                   180
   Val Leu Tyr His Val Glu Glu Leu Asp Pro Ser Ala Leu Ala Asn Phe Pro Leu
   Leu Leu Leu Tyr Val Glu Glu Phe Asp Ala Ser Leu Leu Thr Ser Phe Pro Leu

                                   189                                   198
   Leu Lys Ala Leu Arg Thr Arg Val Ser Asn Leu Pro Thr Val Lys Lys Phe Leu
   Leu Lys Ala Phe Lys Ser Arg Ile Ser Ser Leu Pro Asn Val Lys Lys Phe Leu

                                   207                                   216
   Gln Pro Gly Ser Gln Arg Lys Pro Leu Glu Asp Glu Lys Cys Val Glu Ser Ala
   Gln Pro Gly Ser Gln Arg Lys Pro Ala Met Asp Ala Lys Gln Ile Glu Glu Ala

   Val Lys Ile Phe Ser
   Arg Lys Val Phe Lys Phe
```

**Fig. 2.** Deduced amino acid sequence of the Ya and Yc subunits. Underlined amino acids are divergent residues between the two sequences.

The existence of two classes of cDNA clones is not due to cloning artifacts or DNA sequencing mistakes. Numerous laboratories have constructed cDNA clones corresponding to one or the other class of Ya clones.[17-22]

There is also evidence for microheterogeneity in Ya subunits at the protein level. Wang et al.[26] have prepared monoclonal antibodies to the glutathione S-transferase that recognize a Ya subunit in a Ya dimer but not the Ya subunit in a Ya–Yc heterodimer. These data suggest the monoclonal is recognizing an epitope that is specific for the Ya dimer.

## B. NUCLEOTIDE SEQUENCE ANALYSIS OF THE RAT LIVER GLUTATHIONE S-TRANSFERASE $Yb_1$ AND $Yb_2$ cDNAs AND AMINO ACID SEQUENCES OF THE $Yb_1$ AND $Yb_2$ SUBUNITS

Full-length or nearly full-length cDNA clones complementary to the $Yb_1$ and $Yb_2$ mRNAs have also been constructed and characterized.[27-30] These cDNAs are more related to each other than are the Ya and Yc clones. Over the protein coding region of the $Yb_1$ and $Yb_2$ clones, there is an 84% nucleotide sequence identity; whereas in the 3' untranslated region, there is only a 32% sequence identity. Since the 5' untranslated region of the $Yb_2$ mRNA is not represented in the cDNA clone, the degree of homology between the $Yb_1$ and $Yb_2$ mRNAs in this region is unknown. The nucleotide sequences of $Yb_1$ and $Yb_2$ cDNA clones are presented in Fig. 3. The amino acid sequence of the $Yb_2$ subunit has also been determined directly by conventional protein-sequencing procedures.[31] The sequence of the purified protein agrees with that predicted from nucleotide-sequencing analysis of cDNA clones.

The amino acid sequences of the $Yb_1$ and $Yb_2$ subunits also share significant sequence homology. There is an 80% amino acid sequence identity between the two subunits. The amino acid sequences deduced from nucleotide sequence analysis of the $Yb_1$ and $Yb_2$ subunits are presented in Fig. 4. Lai et al.[29] have published a sequence of a second $Yb_1$ cDNA clone which differs from pGTA/C44 by four nucleotides in the coding region resulting in two amino acid differences. The 5' and 3' untranslated regions of the two $Yb_1$ cDNA clones are identical. At this time it is unclear whether the two $Yb_1$ clones represent expression of two different genes or allelic polymorphism in the outbred rat strains that were used to construct the cDNA clones. However, based on the degree of sequence homology, these different $Yb_1$ cDNAs are most likely due to allelic polymorphism.

```
                                           28                                      55
pGTA/C44 G CTG AAG CCA AAT TGA GAA GAC CAC AGC GCC AGA ACC [ATG] CCT ATG ATA CTG GGA


                                           82                                     109
pGTA/C44 TAC TGG AAC GTC CGC GGG CTG ACA CAC CCG ATC CGC CTG CTC CTG GAA TAC ACA
pGTA/C48                                                                           --


                                          136                                     163
         GAC TCA AGC TAT GAG GAG AAG AGA TAC GCC ATG GGC GAC GCT CCC GAC TAT GAC
         --- A-- --- --- --- --C --- -AG --- AG- --- --G --T --- --- --- --- ---


                                          190                                     217
         AGA AGC CAG TGG CTG AAT GAG AAG TTC AAA CTG GGC CTG GAC TTC CCC AAT CTG
         --- --- --- --- --- -G- --- --- --- --- --- --- --- --- --- --- --- ---


                                          244                                     271
         CCC TAC TTA ATT GAT GGA TCG CGC AAG ATT ACC CAG AGC AAT GCC ATA ATG CGC
         --- --- --- --- --- --G --A -A- --- --C --- --- --- --- --- --C C-- ---


                                          298                                     325
         TAC CTT GCC CGC AAG CAC CAC CTG TGT GGA GAG ACA GAG GAG GAG CGG ATT CGT
         --- --- -G- --G --- --- A-- --T --- --G --- --- --- --- --- A-- --- ---


                                          352                                     379
         GCC GAC ATT GTG GAG AAC CAG GTC ATG GAC AAC CGC ATG CAG CTC ATC ATG CTT
         -TG --- G-- T-- --- --- --A -CT --- --- -C- --- C-A --- T-G GC- --- G-C


                                          406                                     433
         TGT TAC AAC CCC GAC TTT GAG AAG CAG AAG CCA GAG TTC TTG AAG ACC ATC CCT
         --C --- -G- --T --- --- --- -GA A-- --- --- --- -A- --A G-- GGT C-- ---


                                          460                                     487
         GAG AAG ATG AAG CTC TAC TCT GAG TTC CTG GGC AAG CGA CCA TGG TTT GCA GGG
         --- --- --- --- --T --- --C --A --- --- --- --- -AG --- --- --- --- ---


                                          514                                     541
         GAC AAG GTC ACC TAT GTG GAT TTC CTT GCT TAT GAC ATT CTT GAC CAG TAC CAC
         A-- --- A-T --G --- --- --- --T --- -T- --C --T G-C --- --T --A C-- -GT


                                          568                                     595
         AAT TTT GAG CCC AAG TGC CTG GAC GCC TTC CCA AAC CTG AAG GAC TTC CTG GCC
         -TA --- --A --- --- --- --- --- --- --- --- --- --- --- --- G-- --T


                                          622                                     649
         CGC TTT GAG GGC CTG AAG AAG ATC TCT GCC TAC ATG AAT TGC AGC CGC TAC CTC
         --G --- --- --- --- --- --- --A --- -A- --- --- --G A-- G-- --- -T- ---


                                          676                                     703
         TCA ACA CCT ATA TTT TCG AAG TTG GCC CAA TGG AGT AAC AAG TAG GCC CTT GCT
         --C -AG --A --C --- G-A --- A-- --- TTT --- -AC CCA --- --- CA- -AC AAA


                                          730                                     757
         ACA CTG GCA CTC ACA GAG AGG ACC TGT CAA CAT TGG ATC CTG CAG GCA CCC TGG
         GTC -A- A-C TGG GG- T-C TCA TGA GTG -CC TGC --- C-G TG- GCC TAG AG- AT-


                                          784                                     811
         CCT TCT GCA CTG TGG TTC TCT CTC CTT CCT GCT CCC TTC TCC AGC TTT GTC AGC
         G-- CTG --G -CC ACC ACA -GC AG- T-- -TC CTC -TT -C- ATT CC- -G- TC- TC-


                                          838                                     865
         CCC ATC TCC TCA ACC TCA CCC CAG TCA TGC CCA CAT AGT CTT CAT TCT CCC CAC
         AT- TC- --T --C CAG C-C TTG -CT CAG -CA AGC -TC --- TCC TTG GTC T-T -CA


                                          892                                     919
         TTT CTT TCA TAG TGG TCC CCT TCT TTA TTG ACA CCT TAA CAC AAC CTC ACA GTC
         --- --- CAT --- -CC C-T --C -TG C-- C-C TGC ATC --A CC- T-- C-T CA-


                                          946                                     973
         CTT TTC TGT GAT TTG AGG TCT GCC CTG AAC TCA GTC TCC CTA GAC TTA CCC CAA
         TGA --T -CG --G GAC T-T A-C AGA -CC CTG AAT


                                         1000                                    1027
         ATG TAA CAC TGT CTC AGT GCC AGC CTG TTC CTG GTG GGG GAG CTG CCC CAG GCC


         TGT CTC ATC TT
```

**Fig. 3.** DNA sequence analysis of pGTA/C44 (Yb$_1$ clone) and pGTA/C48 (Yb$_2$ clone). Dashed lines in the pGTA/C48 sequence represents nucleotides that are identical to those in the pGTA/C44 sequence.

Fig. 4. Deduced amino acid sequence of the Yb₁ and Yb₂ subunits. Underlined amino acids are divergent residues between the two sequences.

```
                              10                           20
Yb₁   MET Pro MET Ile Leu Gly Tyr Trp Asn Val Arg Gly Leu Thr His Pro Ile Arg Leu Leu
Yb₂   Pro Met Thr Leu Gly Tyr Trp Asp Ile Arg Gly Leu Ala His Ala Ile Arg Leu Phe

                              30                           40
Yb₁   Leu Glu Tyr Thr Asp Ser Ser Tyr Glu Glu Lys Arg Tyr Ala MET Gly Asp Ala Pro Asp
Yb₂   Leu Glu Tyr Thr Asp Thr Ser Tyr Glu Glu Lys Asp Lys Ser Met Gly Asp Ala Pro Asp

                              50                           60
Yb₁   Tyr Asp Arg Ser Gln Trp Leu Asn Glu Lys Phe Lys Leu Gly Leu Asp Phe Pro Asn Leu
Yb₂   Tyr Asp Arg Ser Gln Trp Leu Ser Glu Lys Phe Lys Leu Gly Leu Asp Phe Pro Asn Leu

                              70                           80
Yb₁   Pro Tyr Leu Ile Asp Gly Ser Arg Lys Ile Thr Gln Ser Asn Ala Ile MET Arg Tyr Leu
Yb₂   Pro Tyr Leu Ile Asp Gly Ser His Lys Ile Thr Gln Ser Asn Ala Ile Leu Arg Tyr Leu

                              90                           100
Yb₁   Ala Arg Lys His His Leu Cys Gly Glu Thr Glu Glu Arg Ile Arg Ala Asp Ile Val
Yb₂   Gly Arg Lys His His Asn Leu Cys Gly Glu Thr Glu Glu Arg Ile Arg Ile Asp Val Leu

                              110                          120
Yb₁   Glu Asn Gln Val MET Asp Asn Arg MET Gln Leu Ile MET Leu Cys Tyr Asn Pro Asp Phe
Yb₂   Glu Asn Gln Ala Met Asp Thr Arg Leu Gln Leu Ala Met Val Cys Tyr Ser Pro Asp Phe

                              130                          140
Yb₁   Glu Lys Gln Lys Pro Glu Phe Leu Lys Thr Ile Pro Glu Lys MET Lys Leu Tyr Ser Glu
Yb₂   Glu Lys Lys Lys Pro Glu Tyr Leu Glu Gly Leu Pro Glu Met Lys Leu Tyr Ser Glu

                              150                          160
Yb₁   Phe Leu Gly Lys Arg Pro Trp Phe Ala Gly Asp Lys Val Thr Tyr Val Asp Phe Leu Ala
Yb₂   Phe Leu Gly Lys Gln Pro Trp Phe Ala Gly Asn Lys Ile Thr Tyr Val Asp Phe Leu Val

                              170                          180
Yb₁   Tyr Asp Ile Leu Asp Tyr His Ile Phe Glu Pro Lys Cys Leu Asp Ala Phe Pro Asn
Yb₂   Tyr Asp Val Leu Asp Gln His Arg Ile Phe Glu Pro Lys Cys Leu Asp Ala Phe Pro Asn

                              190                          200
Yb₁   Leu Lys Asp Phe Leu Ala Arg Phe Glu Gly Leu Lys Lys Ile Ser Ala Tyr MET Asn Cys
Yb₂   Leu Lys Asp Phe Val Ala Arg Phe Glu Gly Leu Lys Lys Ile Ser Asp Tyr Met Lys Ser

                              210
Yb₁   Ser Arg Tyr Leu Ser Thr Pro Ile Phe Ser Lys Leu Ala Gln Trp Ser Asn Lys
Yb₂   Gly Arg Phe Leu Ser Pro Ile Phe Ala Phe Met Ala Phe Trp Ala Phe Trp Pro Lys
```

## C. NUCLEOTIDE SEQUENCE ANALYSIS OF A RAT LIVER GLUTATHIONE *S*-TRANSFERASE *P* cDNA CLONE AND THE AMINO ACID SEQUENCE OF THE Y$_p$ SUBUNIT

Suguoka *et al.*[32] has isolated a cDNA clone corresponding to glutathione *S*-transferase P from a cDNA library prepared from poly(A$^+$)-RNA isolated from 2-acetylaminofluorene-induced rat hepatocellular carcinoma. The cDNA insert was 743 nucleotides long and contained an open reading frame of 630 nucleotides encoding 210 amino acids. The deduced molecular weight of glutathione *S*-transferase P is 23,307. A comparison of the GST-P subunit with the rat liver Ya and Yc subunits revealed an overall sequence identity of 32% with the homologous amino acids clustered in three regions of the glutathione *S*-transferase P amino acid sequence. Pemble *et al.*[33] have also isolated an identical clone, pGSTr7, from a *λ*gt 10 library that was constructed from *NN*-dimethyl-4-aminoazobenzene-induced rat hepatoma.

## D. DNA SEQUENCE ANALYSIS OF A HUMAN LIVER cDNA CLONE AND THE COMPLETE AMINO ACID SEQUENCE OF THE CORRESPONDING PROTEIN

Tu and Qian[34] have reported recently the nucleotide sequence of a human glutathione *S*-transferase cDNA clone, pGTH1, which is 810 nucleotides long and contains 66 nucleotides of the 5′ untranslated region and the complete 78 bp of the 3′ noncoding region. The open reading frame is 666 nucleotides long encoding a polypeptide comprising 222 amino acids. The nucleotide sequence of the coding region of the human clone is approximately 80% identical to the rat liver Ya and Yc clones. The predicted amino acid sequence of the human transferase protein is 75% identical to the rat liver Ya and Yc subunits. A comparison of the human cDNA sequence with rat liver Yb subunits revealed sequence homology near the NH$_2$-terminal region (amino acid residues 70–95). Through computer analysis of this region, Tu and Qian[34] have found residues 70–95 are conserved in all the glutathione *S*-transferase subunits. Amino acids 70–91 of the Ya subunit are encoded by exon 4 of the rat liver Ya structural gene.[35] Interestingly, when one compares the sequence identity of amino acids encoded by exon 4 with the comparable region in the Yc subunit, this amino acid region is the most highly conserved throughout the two subunits.[35] Therefore, it is likely that amino acids 70–91 encode domains of the glutathione *S*-transferases, which impart similar functional properties (e.g. GSH binding site).

### III. Structural Analysis of Rat Liver Glutathione *S*-Transferase Genes

## A.  GLUTATHIONE *S*-TRANSFERASE $Y_a$–$Y_c$ GENE FAMILY

Southern blot analysis of genomic DNA using 5' or 3' regions of a Ya clone (pGTB38) or a Yc clone (pGTB42) suggested the presence of at least five Ya genes and two Yc genes in the rat genome.[25] The presence of multiple genes in this family was confirmed by the isolation of four unique genomic fragments from a rat genomic library.[25] One of these genomic clones, λGTB45-15, hybridized to the 5' and 3' ends of pGTB38, the Ya clone, and consequently was a candidate for a full-length genomic clone.[35] The Ya structural gene spans approximately 11 Kb and is comprised of seven exons separated by six introns (Fig. 5). The sizes of the exons and introns in the Ya structural gene is presented in Table 1. Exon 1 is 43 bp in length and encodes most of the 5' untranslated region of the Ya mRNA. Exon 2 is 109 bp in length and encodes amino acids 1–29 of the Ya subunit and 22 bp of the 5' untranslated region. Exon 3 is 52 bp and encodes amino acids 30–46 of the Ya protein. Exons 4 and 5 are 133 bp and 142 bp in length and encode amino acids 47–91 and 92–139, respectively. Exon 6 is 132 bp in length and encodes amino acids 140–183 of the Ya polypeptide whereas exon 7 encodes amino acids 184–222 of the Ya polypeptide and 121 bp of the 3' untranslated region of the Ya mRNA.

Interestingly, the amino acid sequence encoded by exon 3 represents the region that is the most divergent between the Ya and Yc polypeptides. There is only a 36% amino acid sequence identity between the Ya and Yc polypeptide in this region, despite an overall sequence identity of 66% between the two polypeptides. Exons 2 and 4 encode amino acid residues of

TABLE 1

Rat liver glutathione *S*-transferase Ya exon and intron sizes

| Exon | Size determined from DNA sequence analysis base pairs | Intron | Size determined by restriction endonuclease mapping base pairs |
|------|------|------|------|
| 1 | 43  | 1 | 2350 |
| 2 | 109 | 2 | 3500 |
| 3 | 52  | 3 | 650  |
| 4 | 133 | 4 | 2100 |
| 5 | 142 | 5 | 1500 |
| 6 | 132 | 6 | 800  |
| 7 | 234 |   |      |

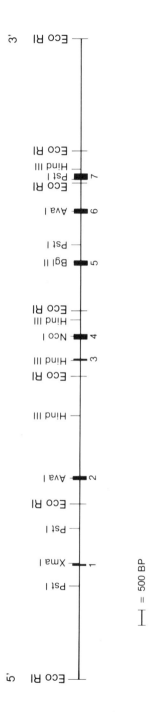

**Fig. 5.** Exon-intron structure of the glutathione *S*-transferase Ya gene. Black boxes represent exons.

the Ya subunit that are most highly conserved in the Yc polypeptide. The amino acid sequences of the Ya polypeptide encoded by exons 2 and 4 are 86% and 91% identical, respectively. Although the significance of the conserved and divergent exons is unclear, it is tempting to speculate that the conserved exons, 2 and 4, encode domains of the subunits that impart identical functional properties such as the glutathione binding site. Exons 3 and 5, which encode regions of the Ya subunit that are very divergent from the Yc subunit, may be important in imparting the substrate specificities to the two subunits. The Ya homodimers have good activity towards $\Delta^5$-andro-stene-3,17 dione whereas the Yc subunit has good activity towards cumene hydroperoxide.

Finally, the precise number of Ya and Yc genes remains unknown. However, Southern blot analysis using 5' and 3' regions of a Ya cDNA clone, pGTB38, indicates the presence of approximately five Ya genes and two Yc genes. Whether all of these genes are expressed in liver or whether some of these genes represent pseudogenes is unknown. To date no laboratory has characterized a full-length glutathione S-transferase Yc structural gene.

## B. GLUTATHIONE S-TRANSFERASE P GENE

Okuda et al.[36] has isolated the glutathione S-transferase P structural gene from a rat liver genomic library. The rat glutathione S-transferase P gene is about 3000 bp long and like the Ya gene contains seven exons separated by six introns. The start of transcription maps 70 nucleotides upstream from the translation initiation site. Okuda et al.[36] have determined the complete nucleotide sequence of the genomic clone and have demonstrated it corresponds in sequence to the cDNA clone, pGP5, described earlier. In addition to the full-length structural gene, Okuda et al.[36] isolated several processed-type pseudogenes, which originated most likely by reverse transcription followed by insertion at specific sites.

## C. GLUTATHIONE S-TRANSFERASE Yb GENE FAMILY

Southern blots of rat liver genomic DNA with a Yb probe also indicate a multigene family encoding these glutathione S-transferases.[29] Tu et al.[37] have recently determined the exon/intron structure of a glutathione S-transferase Yb$_2$ gene, which spans approximately 5 Kb. The gene is split into eight exons encoding 12, 25-1/3, 21-2/3, 27, 34, 32, 37 and 29 amino acids, respectively. The first and eight exons contain the 5' and 3' untranslated regions, respectively.

## IV. Regulation of Rat Liver Glutathione *S*-Transferase Genes by Xenobiotics

Early studies indicated that various glutathione *S*-transferase isozymes were elevated in the livers of rats administered 3-methylcholanthrene and phenobarbital. The increase in enzyme content and/or activity was paralleled by an elevation in the translational activity of the Ya mRNA.[38,39] Using *in vitro* translation systems and immunoprecipitation with specific antibodies raised against purified glutathione *S*-transferase Ya–Yc hetero-dimer, an increase in the translational activity of the Ya mRNA could be detected as early as 4 h after a single administration of phenobarbital, reaching maximal induction (~five- to seven-fold) between 16 and 24 h.[40] RNA blot analysis using cDNA clones complementary to the Ya mRNA confirmed that the increase in the translational activity of the Ya mRNA was due to an accumulation in the steady-state level of mRNA specific for the Ya subunit (ref. 19 and Table 2). The Yc mRNA was only modestly affected (~1·5-fold) by phenobarbital (Fig. 6 and Table 2) or 3-methylcholanthrene treatment. The use of specific regions of the Yb$_1$ and Yb$_2$ cDNA clones in RNA slot blots demonstrated that both of these mRNAs are elevated by phenobarbital (Figs. 7 and 8, Table 2) and 3-methylcholanthrene.[19] The accumulation of glutathione *S*-transferase mRNAs in response to xenobiotic administration prompted our laboratory to examine whether the genes encoding the glutathione *S*-transferases were transcriptionally activated.[41] Consequently, nuclei were isolated from the livers of rats treated for various

TABLE 2

Effect of phenobarbital on the level of rat liver
glutathione *S*-transferase mRNAs

| Hours after phenobarbital administration | Relative mRNA levels | | | |
|:---:|:---:|:---:|:---:|:---:|
| | Ya | Yb$_1$ | Yb$_2$ | Yc |
| 0 | 1 | 1 | 1 | 1 |
| 4 | — | 2·7 | 2·7 | 1 |
| 8 | — | 2·8 | 4·2 | 1 |
| 12 | 5·4 | — | — | — |
| 16 | — | 3·6 | 4·9 | 2·0 |
| 24 | 7·9 | 5·8 | 4·7 | 1·6 |

The relative level of rat liver glutathione *S*-transferase mRNAs were determined by RNA blots using specific cDNA probes.

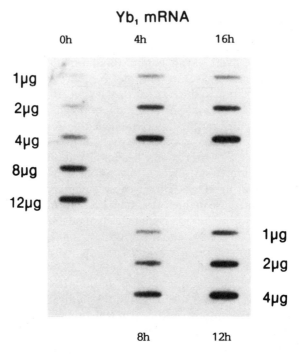

**Fig. 6.** Effect of phenobarbital on the level of glutathione *S*-transferase Yb₁ mRNA. Phenobarbital (80 mg/kg) was injected (i.p.) into rats and total RNA was isolated from their livers at the indicated times. Poly(A⁺)-RNA was isolated from the total RNA and pipetted onto nitrocellulose filter paper. The RNA was hybridized with the 3′ end of clone pGTA/C44, a Yb₁-specific probe.

times with phenobarbital and 3-methylcholanthrene. Isolated nuclei were allowed to complete nascent RNA chains *in vitro* using $^{32}$P UTP as the radiolabel. $^{32}$P-RNA was isolated and hybridized to various glutathione *S*-transferase cDNA clones. The amount of $^{32}$P-RNA that hybridized reflects the transcriptional activity of the glutathione *S*-transferase genes. The transcriptional rate of the Ya genes is elevated eight-fold at 16 h after phenobarbital administration, whereas the Yb genes were elevated five-fold at 6 h after the administration of this xenobiotic (ref. 41 and Table 3).

In contrast, the transcriptional rate of the Ya genes is elevated eight-fold at 16 h after 3-methylcholanthrene administration, whereas the Yb genes were elevated five-fold at 6 h after the administration of this xenobiotic (ref. 41 and Table 3). The extent of transcriptional activity is sufficient to account for the accumulation of steady-state mRNA after xenobiotic administration.

The mechanism(s) by which these various xenobiotics activate transcriptionally the glutathione *S*-transferase genes has not been resolved. Our laboratory (Telakowski-Hopkins, King and Pickett, unpublished) has

## Yb₂ mRNA

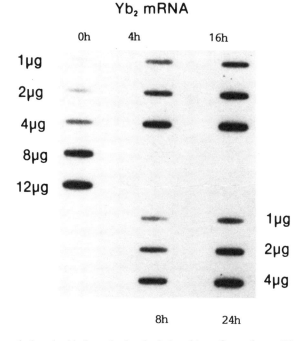

**Fig. 7.** Effect of phenobarbital on the level of glutathione S-transferase Yb₂ mRNA. This experiment is identical to the one presented in Fig. 7; however, the 3' end of pGTA/c48 was used as a specific Yb₂ probe.

TABLE 3

Transcriptional activity of the glutathione S-transferase
genes in response to 3-methylcholanthrene
administration

| Hours after 3-methylcholanthrene administration | Gene transcription (ppm) | |
|---|---|---|
| | Ya gene | Yb gene |
| 0 | 5 | 3 |
| 2 | 10 | 8 |
| 4 | 12 | 8·2 |
| 6 | 8 | 12 |
| 8 | 22 | 3 |
| 12 | 30 | 4 |
| 16 | 42 | 5 |

Parts per million hybridized was calculated based on the total cpm incubated with each cDNA filter.

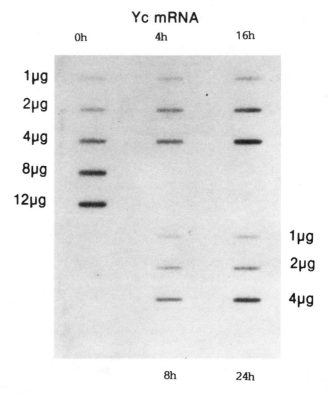

**Fig. 8.** Effect of phenobarbital on the level of glutathione *S*-transferase Yc mRNA. This experiment is identical to the one presented in Fig. 7; however, the 3′ end of pGTB42 was used as a specific Yc probe.

recently constructed chimeric genes using 1·6 kb of the 5′ flanking region of a glutathione *S*-transferase Ya gene fused to the structural gene encoding chloramphenicol acetyltransferase (CAT). We have transfected this chimeric gene into rat, mouse and human hepatoma cell lines and have found that the 1·6 Kb structural gene fragment drives CAT activity, which indicates that the fragment contains a functional glutathione *S*-transferase promotor (Fig. 9). When cells containing the glutathione *S*-transferase-CAT structural gene were treated with β-naphthoflavone, a compound which binds to the AH receptor, CAT activity is elevated three- to seven-fold in the hepatoma cell lines. These data suggest that the glutathione *S*-transferase Ya structural gene contains *cis*-acting regulatory elements that are responsible for transcriptional activation. If a chimeric gene containing CAT under the control of the SV40 promoter and enhancer is transfected into the cell

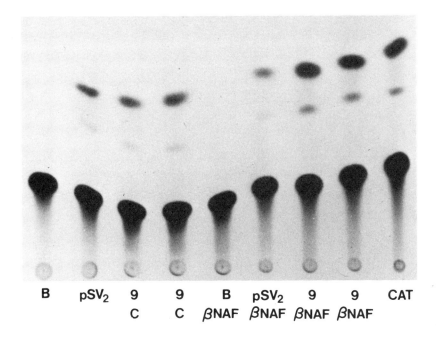

| B | pSV₂ | 9 | 9 | B | pSV₂ | 9 | 9 | CAT |
|---|------|---|---|---|------|---|---|-----|
|   |      | C | C | βNAF | βNAF | βNAF | βNAF |   |

**Fig. 9.** Effect of $\beta$-naphthoflavone on the activity of chloramphenicol acetyl transferase (CAT) activity in a rat hepatoma cell line transfected with chimeric plasmids containing the glutathione $S$-transferase 5′ flanking region fused to the CAT structural gene. (B): transfected hepatoma cells; (pSV₂); hepatoma cells transfected with CAT structural gene driven by the SV40 promoter and enhancer; (9C): hepatoma cells transfected with CAT structural gene driven by the 5′ flanking region of the glutathione $S$-transferase Ya structural gene; (B-$\beta$NAF): un-transfected hepatoma cells treated with 50 $\mu$M $\beta$-naphthoflavone; (pSV₂ $\beta$NAF): identical to the pSV₂ experiment, however cells were treated with $\beta$-naphthoflavone; (9$\beta$NAF): identical to the 9C experiment; however, cells were treated with $\beta$-naphthoflavone; (CAT): control assay using purified chloramphenicol acetyltransferase.

In the CAT enzymatic assay $^{14}$C-chloramphenicol is converted to the acetylated forms, which migrate at a faster rate on TLC plates.

line, no increase in CAT activity occurs when $\beta$-naphthoflavone is added. The trans-acting factor(s), which presumably bind to the *cis*-acting regulatory regions, also appears to be highly conserved in rodent and human cells. As mentioned previously, CAT activity is increased in the presence of $\beta$-naphthoflavone in mouse, rat and human hepatoma cells transfected with the transferase-CAT construct.

At the present time it is unclear whether the AH receptor is directly or indirectly involved in the induction of the glutathione $S$-transferase genes. However, in the P-450 system, Whitlock and colleagues[42–44] have demonstrated quite eloquently that the gene encoding the major 3-methyl-

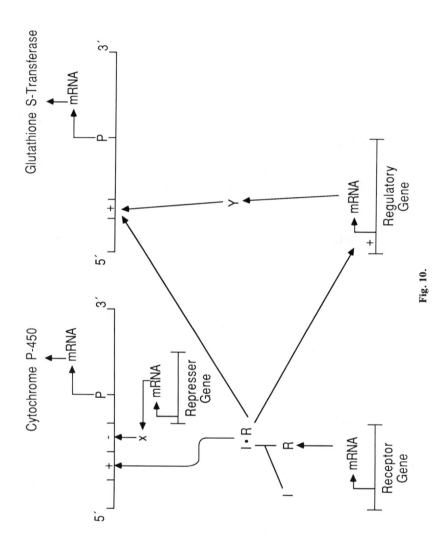

Fig. 10.

cholanthrene inducible P-450 isoenzyme ($P_1$-450) contains a core promoter, TCDD or dioxin-responsible elements and a negative regulatory element that interacts with a repressor protein. Similar findings have also been obtained by Fujisawa-Sehara *et al.*[45] and Gonzalez and Nebert.[46] Sogawa *et al.*[47] have located similar responsive elements on the rat cytochrome P-450c gene and have deduced a consensus sequence

$$5'\text{-} \begin{matrix} G \\ C \end{matrix} N \begin{matrix} T \\ G \end{matrix} \begin{matrix} A \\ G \end{matrix} G\ C\ T\ G\ G\ G\ 3'$$

which presumably forms part of the dioxin-responsive element. It is presently unknown whether the glutathione *S*-transferase Ya structural gene is directly or indirectly regulated by the dioxin receptor. A possible scheme for the regulation of glutathione *S*-transferase Ya and $P_1$-450 gene expression is presented in Fig. 10. In this model the dioxin receptor is shown to interact with ligand (l) and regulate glutathione *S*-transferase gene expression directly or interact with the ligand and activate the transcription of a regulatory gene producing a *trans*-activating protein ($y$) that is responsible for activation.

## V. Expression of Glutathione *S*-Transferase Genes and Genes Encoding Other Drug Metabolizing Enzymes During Chemical Carcinogenesis

The administration of a single dose of diethylnitrosamine to rats followed by three consecutive doses of 2-acetyl aminofluorene plus a partial hepatectomy induces foci of altered hepatocytes, which have been referred to as persistent hepatocyte nodules, hyperplastic nodules, and preneoplastic nodules.[48,49] These nodules are thought to serve as precursors to hepatocellular carcinomas and have been used as a model system to study chemically-induced neoplastic transformation. Biochemically the nodules have elevated

---

**Fig. 10.** A model to account for the regulation of the glutathione *S*-transferase gene expression is presented along with the current model governing $P_1$-450 gene regulation by polycyclic aromatic hydrocarbons. In this model the AH or dioxin receptor (R) binds polycyclic aromatic hydrocarbons (l) forming a receptor ligand complex that is translocated to the nucleus and interacts with positive regulatory elements (polycyclic aromatic hydrocarbon responsive elements) leading to the transcriptional activation of the $P_1$-450 gene and possibly the glutathione *S*-transferase Ya gene. Alternatively the receptor ligand complex (I.R.) could potentially interact with a regulatory gene leading to the transcriptional activation of the regulatory gene producing a trans-acting protein, which regulates expression of the glutathione *S*-transferase Ya gene. To date, we have not been able to distinguish between these two possibilities. Finally, in the P-450 system a repressor protein (X) is though to regulate negatively the constitutive expression of the $P_1$-450 genes in the absence of ligand.

levels of various phase II drug-metabolizing enzymes (e.g. glutathione S-transferases, NAD(P)H: quinone reductase and epoxide hydrolase); however, cytochrome P-450 levels are depressed.[50–56] The elevated levels of the phase II drug-metabolizing enzymes is consistent with the observation that persistent hepatocyte nodules are more resistant to the cytotoxic effects of carcinogens as compared to normal liver.

In order to understand the mechanisms of chemical carcinogenesis, a basic understanding of the regulation of the enzymes involved in carcinogen metabolism is essential. Since cDNA clones complementary to various glutathione S-transferase mRNAs as well as mRNAs specific for other drug metabolizing enzymes have been constructed, these cDNA clones can be used as probes to elucidate the molecular basis for the observed changes in protein level. We have found that the glutathione S-transferase Ya and Yb mRNAs and quinone reductase mRNA are elevated 2·8-fold, 5·3-fold and 10-fold respectively, in nodule tissue compared to surrounding liver tissue or normal liver (ref. 57 and Table 4). If animals have undergone nodule induction and then are treated with 3-methylcholanthrene, quinone reductase mRNA and glutathione S-transferase Yb mRNA are elevated to levels that exceed their level in nodules or after 3-methylcholanthrene administration alone.[57,58] The quinone reductase mRNA level was elevated 20-fold over the level in untreated rats whereas the Yb mRNA was elevated 13-fold over the level observed in untreated rats (Table 5). However, the Ya mRNA level in liver could not be elevated further after 3-methylcholanthrene administration to rats that had hepatic nodules (Table 5). These data indicate that the regulation of the Ya mRNA in persistent hepatocyte nodules is aberrant; however, the precise defect in the regulatory process has not been delineated.

TABLE 4

Glutathione S-transferase (Ya/Yc and Yb) and DT-diaphorase mRNA levels in normal liver tissue, surrounding liver tissue and nodular tissue

| Tissue | Relative mRNA level | | |
|---|---|---|---|
| | DT-Diaphorase | Ya/Yc mRNA | Yb mRNA |
| Normal liver | 1·0 | 1·0 | 1·0 |
| Surrounding liver | 1·7 | 1·2 | 1·8 |
| Nodules | 7·0 | 2·8 | 5·3 |

RNA levels were determined by scanning X-ray films obtained from RNA blots or Northern blots.

Suguoka *et al.*[32] have demonstrated that the glutathione *S*-transferase P mRNA (placental form) was substantially elevated in acetyl aminofluorene-induced hepatocellular carcinoma and Satoh *et al.*[12] have demonstrated that the transferase P-protein is markedly elevated in preneoplastic nodules. Pemble *et al.*[33] have also demonstrated an elevation in transferase P (subunit 7) mRNA in *NN*-dimethyl-4-aminoazobenzene (DAB) induced rat hepatomas.

In an attempt to correlate the elevation in mRNA with changes at the DNA level, we examined whether the quinone reductase or the epoxide hydrolase gene was amplified in the nodules. No amplification of these genes was observed. Furthermore, neither the epoxide hydrolase nor the quinone reductase genes were rearranged to any significant extent. Finally, we examined the extent of hypomethylation of the quinone reductase and epoxide hydrolase genes using two enzymes, Msp I and Hpa II. These two enzymes recognize and cleave the sequence CCGG but Hpa II will not cleave the sequence if the internal cytosine is methylated. Consequently, one can assess the degree of hypomethylation of a gene by digesting genomic DNA with Msp I or Hpa II, electrophoresizing the DNA on agarose gels, blotting the digested DNA to nitrocellulose paper and probing with radiolabelled cDNA. The results from these experiments indicated that the DT-diaphorase gene was hypomethylated in nodular tissue as compared to the gene in surrounding liver or normal liver.[58]

In work from our laboratory (V. D.-H. Ding, and C. B. Pickett, unpublished), we have also examined the methylation status of the epoxide hydrolase gene. Similar to DT-diaphorase, the epoxide hydrolase gene is hypomethylated in nodular tissue compared to normal tissue. In the

TABLE 5

Glutathione *S*-transferase (Ya/Yc and Yb) and DT-diaphorase mRNA levels in normal liver tissue and in surrounding liver tissue and nodular tissue from 3-methylcholanthrene-treated rats

| Treatment | Tissue | Relative mRNA level | | |
|---|---|---|---|---|
| | | DT-Diaphorase | Ya/Yc | Yb |
| None | Untreated liver | 1 | 1 | 1 |
| 3-Methylcholanthrene | Normal liver | 6 | 4·3 | 3 |
| 3-Methylcholanthrene plus nodule induction | Surrounding liver | 6 | 2·9 | 4·6 |
| 3-Methylcholanthrene plus nodule induction | Nodules | 21 | 3·2 | 13 |

hepatomas that develop in the rats, the extent of hypomethylation of the epoxide hydrolase gene is maintained as are the elevated mRNA levels. Therefore, the hypomethylation of the epoxide hydrolase and quinone reductase gene is a persistent change that occurs early during chemical carcinogenesis. The finding that hypomethylation of these two genes occurs is consistent with *in vitro* data, which indicates that various chemical carcinogens inhibit DNA methylase activity. Lapeyre and Becker[59] have demonstrated a significant decrease in methylation of DNAs isolated from nodules and hepatocellular carcinomas induced by 2-acetylaminofluorene. Pfohl-Leszkowicz *et al.*[60] have detected binding of 2-acetylaminofluorene to C-8 of guanine. The binding of 2-acetylaminofluorene to DNA inhibited the ability of rat brain DNA methyltransferase to methylate native DNA. Thus, the results of these *in vitro* studies support our finding that hypomethylation of specific gene sequences does occur in nodular tissue.

Cowan *et al.*[61] have recently isolated a doxorubicin (adriamycin)-resistant human breast cancer cell line (DOX$^R$ MCF7 cells) that have developed the phenotype of multidrug resistance. Using a cDNA clone complementary to the Yp subunit, constructed in Muramatsu's laboratory,[32] Cowan *et al.*[61] found a marked elevation in Yp mRNA in the DOX$^R$ cells compared to the wild type. In addition, they also found a decrease in aryl hydrocarbon hydroxylase activity. These findings suggest that at least the biochemical changes in the human breast cancer cells are similar to the biochemical changes observed in persistent hepatocyte nodules induced in rat liver by chemical carcinogenesis.

Finally, Manoharan *et al.*[62] have ligated the coding region of glutathione *S*-transferase Ya cDNA (clone pGTB38) to the SV40 early region promoter or the herpes simplex thymidine kinase gene. One construct containing the SV40 promoter produced significant levels of Ya mRNA and glutathione *S*-transferase Ya–Ya specific enzyme activity ($\Delta^5$-androstene-3,17-dione isomerization) in COS cells. When a mixed population of COS cells containing Ya–Ya over expressing cells was treated with benzo(a)pyrene ($\pm$) anti-diol epoxide, which is a cytotoxic agent and a substrate for the glutathione *S*-transferase Ya homodimer, a 20- to 30-fold enrichment in clones of over-expressing cells was seen among those cells surviving treatment. These data indicate that over-expression of the glutathione *S*-transferase Ya homodimer in mammalian cells is accompanied by significant biological resistance to a known alkylating molecule.

In summary, it appears that over-expression of glutathione *S*-transferases can be correlated with the development within cells of resistance to cytotoxic agents. It remains to be elucidated if this is a general phenomenon. However, it does provide a rational basis for understanding the mechanism underlying the development of acquired resistance to antineoplastic agents.

## VI. Tissue-specific Expression of Rat Liver Glutathione S-Transferase Genes

In 1981, Scully and Mantle[63] reported the absence of a Ya subunit in testis and the absence of Yb subunit in the kidney. A few years earlier, Guthenberg and Mannervik[64] identified a unique glutathione S-transferase from lung that was absent from rat liver. Results such as these suggested that the rat glutathione S-transferases are expressed in a tissue-specific manner. Tu et al.[65] extended these studies and used in vitro translation and protein púrification as tools to begin to elucidate the differences in expression of rat glutathione S-transferases in various tissues. Poly(A$^+$)-RNA were isolated from six rat tissues (heart, kidney, liver, lung, spleen and testis) translated in the rabbit reticulocyte lysate system in the presence of $^{35}$S-methionine. The glutathione S-transferase subunits synthesized in vitro were purified from the translation mixtures by affinity chromatography on S-hexyl glutathione-linked Sepharose 6B columns. The affinity-bound fractions were analysed by SDS/polyacrylamide gel electrophoresis. Using this methodology, Tu et al.[65] found a transferase subunit ($M_r = 22\,000$) that was detected in translation products directed by heart, kidney, lung, spleen and testis; however, this subunit was missing in liver.

Li et al.[66] have demonstrated that in rat brain the liver Ya subunit is not expressed. In addition, it was demonstrated that two transferase subunits of distinct size ($M_r = 26\,300$ and $25\,000$) was expressed in brain. Using a Yb cDNA clone, Li et al.[66] also demonstrated the existence by Northern blot analysis of a mRNA species of ~1300 nucleotides that hybridized to the probe. More recently, Abramovitz and Listowsky[67] have isolated and characterized a novel Yb cDNA (Yb$_3$) clone from a rat brain $\lambda$gt 11 expression library. Using the 3′ untranslated region of the Yb$_3$ clone as a specific probe, Abramovitz and Listowsky observed that the Yb$_3$ mRNA is 1300 bp in length and is expressed primarily in the brain. The deduced amino acid sequence of the Yb$_3$ subunit indicates it shares significant sequence homology to the Yb$_1$ and Yb$_2$ subunits (80%). These data provide the first direct proof for the existence of a third related subunit of the Yb family and that expression of glutathione S-transferase Yb genes does occur in a tissue-specific manner.

Pemble et al.[33] have also demonstrated using a cDNA complementary to glutathione S-transferase P (subunit 7) mRNA that expression of mRNA occurs in a very tissue-specific manner. Namely, mRNA hybridizing to this cDNA probe was detected in epididymis, kidney, adrenal gland, lung, spleen, skeletal muscle and heart but not in testis or normal liver.

Hayes and Mantle[68] have examined the tissue distribution of various glutathione S-transferase subunits by Western blot analysis using polyclonal antibody. Their data provides additional support at the protein level for a significant difference in the tissue-specific expression of glutathione $S$-transferase subunits. It remains to be elucidated what the structural and functional differences are between glutathione $S$-transferases expressed in liver and extrahepatic tissue. These data can only be obtained by a combined enzymological and molecular biological approach.

## VII. Conclusion

This review has attempted to bring the reader up to date on the structure and regulation of mammalian glutathione $S$-transferases. It has focussed mostly on the rat as a model to study gene structure and regulation; however, other mammalian systems will undoubtedly be similar in the number of glutathione $S$-transferase genes and the basic mechanisms that control their expression. It should be possible in the near future to delineate the regulatory elements in the 5' flanking region of the transferase genes that are responsive to various xenobiotics and to isolate the $trans$-acting factors that are required for transcriptional activation.

At present it is unclear whether the glutathione $S$-transferase genes are transcriptionally-activated directly or indirectly by the AH or dioxin receptor. If the transcriptional activation is an indirect mechanism, it suggests that another regulatory gene is activated by polycyclic aromatic hydrocarbons producing a protein which serves as a $trans$-activating protein (Fig. 10).

Although it has been demonstrated that phenobarbital regulates the expression of glutathione $S$-transferase genes in the rat, the availability of phenobarbital-responsive cell lines is critical to define regulatory elements involved in transcriptional activation of the transferase genes. The use of primary hepatocyte cultures may be useful in examining phenobarbital induction.

The role chromatin structure plays in the expression of glutathione $S$-transferase genes is virtually unknown. Differences in nuclease sensitivities may help in identifying important regions of the transferase genes required for developmental regulation or tissue-specific expression.

The role that specific amino acid residues or domains play in substrate binding, ligand binding (e.g. heme, bilirubin) and catalytic activity can now be approached using site-directed mutagenesis. The lack of post-translational modification and the cytosolic localization of the glutathione $S$-transferase should make them ideal candidates for expression in $E. coli$ and X-ray crystallographic analysis. These techniques will be invaluable in correlating various amino acid domains or residues with unique functional properties of the enzymes.

One of the major problems in current chemotherapy is the establishment of resistant cell populations following drug exposure. Many investigators have found an elevation in glutathione S-transferase activity in cell lines resistant to various alkylating agents. The identity of individual glutathione S-transferases involved in the metabolism of these agents and the molecular mechanisms underlying their elevation needs to be established. These studies should elucidate the role that alterations in glutathione S-transferase gene expression plays in acquired resistance to anti-tumor drugs.

## ACKNOWLEDGEMENTS

I would like to thank my research colleagues (Claudia A. Telakowski-Hopkins, Gloria J.-F. Ding, Victor D.-H. Ding, Gail S. Rothkopf, Jacinta B. Williams and Ron G. King) who have contributed to the research in my laboratory. I would also like to thank my collaborators, Anthony Y. H. Lu and Ross G. Cameron for their support. Finally, I would like to acknowledge Joan Kiliyanski for her assistance in the preparation of this essay.

## REFERENCES

1. Litwack, G., Ketterer, B. & Arias, I. M. (1971). Ligandin: A hepatic protein which binds steroids, bilirubin, carcinogens, and a number of exogenous organic anions. *Nature (London)* **234**, 466–467.
2. Arias, I. M., Fleischner, G., Kirsch, R., Mishkin, S. & Gatmaitan, Z. (1976). In *Glutathione: Metabolism and Function* (Arias, I. M. & Jakoby, W. B., eds), pp. 175–188. Raven Press, New York.
3. Jakoby, W. B. & Habig, W. H. (1980). In *Enzymatic Basis of Detoxification* (Jakoby, W. B., ed.), pp. 63–94. Academic Press, London and New York.
4. Bhargava, M. M., Ohmi, N., Listowsky, I. & Arias, I. M. (1980). Subunit composition, organic anion binding, catalytic and immunological properties of ligandin from rat testis. *J. Biol. Chem.* **225**, 724–727.
5. Habig, W. H., Pabst, M. J. & Jakoby, W. B. (1974). Glutathione S-transferases. The first enzymatic step in mercapturic acid formation. *J. Biol. Chem.* **249**, 7130–7139.
6. Habig, W. H., Pabst, M. J. & Jakoby, W. B. (1976). Glutathione S-transferase AA from rat liver. *Arch. Biochem. Biophys.* **175**, 710–716.
7. Mannervik, B. & Jensson, H. (1982). Binary combination of four protein subunits with different catalytic specificities explain the relationship between six basic glutathione S-transferases in rat liver cytosol. *J. Biol. Chem.* **257**, 9909–9912.
8. Homma, H. & Listowsky, I. (1985). Identification of Yb glutathione S-transferase as a major rat liver protein labeled with dexamethasone 21-methane sulfonate. *Proc. Natl. Acad. Sci. USA* **82**, 7165–7169.
9. Bass, N. M., Kirsch, R. E., Tuff, S. A., Marks, I. & Saunders, S. J. (1977). Ligandin heterogeneity: Evidence that the two non-identical subunits are the monomers of two distinct proteins. *Biochem. Biophys. Acta* **492**, 163–175.
10. Reddy, C. C., Li, N.-Q. & Tu, C.-P. D. (1984). Identification of a new glutathione S-transferase from rat liver cytosol. *Biochem. Biophys. Res. Commun.* **121**, 1014–1020.

11. Hayes, J. D. (1984). Purification and characterization of glutathione S-transferases P, S and N. *Biochem. J.* **224**, 839–852.
12. Satoh, K., Kitahara, A., Soma, Y., Inaba, Y., Hatayama, I. & Sato, K. (1985). Purification, induction, and distribution of placental glutathione transferase: A new marker enzyme for preneoplastic cells in the rat chemical hepatocarcinogenesis. *Proc. Natl. Acad. Sci. USA* **82**, 3964–3968.
13. Hayes, J. D. (1986). Purification and physical characterization of glutathione S-transferase K. *Biochem. J.* **233**, 789–798.
14. Jakoby, W. B., Ketterer, B. & Mannervik, B. (1984). Glutathione transferases: Nomenclature. *Biochem. Pharmacol.* **33**, 2539–2540.
15. Mannervik, B. (1985). The isozymes of glutathione transferase. *Adv. Enzym. Rel. Areas Mol. Biol.* **57**, 357–417.
16. Chasseaud, L. F. (1979). The role of glutathione and glutathione S-transferases in the metabolism of chemical carcinogens and other electrophilic agents. *Adv. Cancer Res.* **29**, 175–274.
17. Taylor, J. B., Craig, R. K., Beale, D. & Ketterer, B. (1984). Construction and characterization of a plasmid containing complementary DNA to mRNA encoding the N-terminal amino acid sequence of the rat glutathione S-transferase. *Biochem. J.* **219**, 223–231.
18. Lai, H.-C. J., Li, N.-Q., Weiss, M. J., Reddy, C. C. & Tu, C.-P. D. (1984). The nucleotide sequence of a rat liver glutathione S-transferase subunit cDNA clone. *J. Biol. Chem.* **259**, 5536–5542.
19. Pickett, C. B., Telakowski-Hopkins, C. A., Ding, G. J.-F., Argenbright, L. & Lu, A. Y. H. (1984). Rat liver glutathione S-transferases: Complete nucleotide sequence of a glutathione S-transferase mRNA and the regulation of the Ya, Yb and Yc mRNAs by 3-methylcholanthrene and phenobarbital. *J. Biol. Chem.* **259**, 5182–5188.
20. Daniel, V., Sarid, S., Bar-Nun, S. & Litwack, G. (1983). Rat ligandin mRNA molecular cloning and sequencing. *Arch. Biochem. Biophys.* **227**, 266–271.
21. Kalinyak, J. E. & Taylor, J. M. (1982). Rat glutathione S-transferase: Cloning of double-stranded cDNA and induction of its mRNA. *J. Biol. Chem.* **257**, 523–530.
22. Tu, C.-P. D., Weiss, M. J., Karakawa, W. W. & Reddy, C. C. (1982). Cloning and sequence analysis of a cDNA plasmid for one of the rat liver glutathione S-transferase subunits. *Nucleic Acids Res.* **10**, 5407–5419.
23. Tu, C.-P. D., Lai, H.-C. J., Li, N.-Q., Weiss, M. J. & Reddy, C. C. (1984). The Yc and Ya subunits of rat liver glutathione S-transferases are products of separate genes. *J. Biol. Chem.* **259**, 9434–9439.
24. Telakowski-Hopkins, C. A., Rodkey, J. A., Bennett, C. D., Lu, A. Y. H. & Pickett, C. B. (1985). Rat liver glutathione S-transferases: Construction of a cDNA clone complementary to a Yc mRNA and prediction of the complete amino acid sequence of a Yc subunit. *J. Biol. Chem.* **260**, 5820–5825.
25. Rothkopf, G. S., Telakowski-Hopkins, C. A., Stotish, R. L. & Pickett, C. B. (1986). Multiplicity of glutathione S-transferase genes in the rat and association with a Type 2 Alu repetitive element. *Biochemistry* **25**, 993–1002.
26. Wang, I. Y., Tung, E., Wang, A.-C., Argenbright, L., Wang, R., Pickett, C. B. & Lu, A. Y. H. (1986). Multiple Ya subunits of glutathione S-transferase detected by monoclonal antibodies. *Arch. Biochem. Biophys.* **245**, 543–547.
27. Ding, G. J.-F., Lu, A. Y. H. & Pickett, C. B. (1985). Rat liver glutathione S-transferases: Nucleotide sequence analysis of a $Yb_1$ cDNA clone and predic-

tion of the complete amino acid sequence of the Yb$_1$ subunit. *J. Biol. Chem.* **260**, 13 628–13 271.

28. Ding, G. J.-F., Ding, V. D.-H., Rodkey, J. A., Bennett, C. D., Lu, A. Y. H. & Pickett, C. B. (1986). Rat liver glutathione S-transferases: DNA sequence of a Yb$_2$ cDNA clone and regulation of the Yb$_1$ and Yb$_2$ mRNAs by phenobarbital. *J. Biol. Chem.* **261**, 7952–7957.

29. Lai, H. C.-J., Grove, G. & Tu, C.-P. D. (1986). Cloning and sequence analysis of a cDNA for a rat liver glutathione S-transferase Yb subunit. *Nucleic Acids Res.* **14**, 6101–6114.

30. Lai, H. C.-J. & Tu, C.-P. D. (1986). Rat glutathione S-transferase supergene family. Characterization of an anionic Yb subunit cDNA clone. *J. Biol. Chem.* **261**, 13 793–13 799.

31. Alin, P., Mannervik, B. & Jornvall, H. (1986). Cytosolic rat liver glutathione S-transferase 4-4. Primary structure of the protein reveals extensive differences between homologous glutathione S-transferases of classes Alpha and Mu. *Eur. J. Biochem.* **156**, 343–350.

32. Sugoka, Y., Kano, T., Okuda, A., Sakai, M., Kitagawa, T. & Muramatsu, M. (1985). Cloning and the nucleotide sequence of rat glutathione S-transferase P cDNA. *Nucleic Acids Res.* **13**, 6049–6057.

33. Pemble, S. E., Taylor, J. B. & Ketterer, B. (1986). Tissue distribution of rat glutathione transferase subunit 7, a hepatoma marker. *Biochem. J.* **240**, 885–889.

34. Tu, C.-P. D. & Qian, B. (1986). Human liver glutathione S-transferases: Complete primary sequence of an Ha subunit cDNA. *Biochem. Biophys. Res. Commun.* **141**, 229–237.

35. Telakowski-Hopkins, C. A., Rothkopf, G. S. & Pickett, C. B. (1986). Structural analysis of a rat liver glutathione S-transferase Ya gene. *Proc. Natl. Acad. Sci. USA* **83**, 9393–9397.

36. Okuda, A., Sakai, M. & Muramatsu, M. (1987). The structure of the rat glutathione S-transferase P gene and related pseudogenes. *J. Biol. Chem.* **262**, 3858–3863.

37. Tu, C. P.-D., Lai, H.-C. J. & Reddy, C. C. (1987). The rat glutathione S-transferases supergene family: Molecular basis of gene multiplicity. In *Glutathione S-Transferases and Carcinogenesis* (Mantle, T. J., Pickett, C. B. & Hayes, J. D., eds), pp. 87–110. Taylor and Francis, London.

38. Pickett, C. B., Wells, W., Lu, A. Y. H. & Hales, B. F. (1981). Induction of translationally active rat liver glutathione S-transferase B messenger RNA by phenobarbital. *Biochem. Biophys. Res. Commun.* **99**, 1002–1010.

39. Pickett, C. B., Telakowski-Hopkins, C. A., Donohue, A. M., Lu, A. Y. H. & Hales, B. F. (1982). Differential induction of rat hepatic cytochrome P-448 and glutathione transferase B messenger RNAs by 3-methylcholanthrene. *Biochem. Biophys. Res. Commun.* **104**, 611–619.

40. Pickett, C. B., Donohue, A. M., Lu, A. Y. H. & Hales, B. F. (1982). Rat liver glutathione S-transferase B: The functional mRNAs specific for the Ya–Yc subunits are induced differentially by phenobarbital. *Arch. Biochem. Biophys.* **215**, 539–543.

41. Ding, V. D.-H. & Pickett, C. B. (1985). Transcriptional regulation of rat liver glutathione S-transferase genes by phenobarbital and 3-methylcholanthrene. *Arch. Biochem. Biophys.* **240**, 553–559.

42. Jones, P. B. C., Galeazzi, D. R., Fisher, J. M. & Whitlock, J. P., Jr (1985).

Control of cytochrome $P_1$-450 gene expression by dioxin. *Science* **227**, 1499–1502.

43. Jones, P. B. C., Durrin, L. K., Galeazzi, D. R. & Whitlock, J. P., Jr (1986). Control of cytochrome $P_1$-450 gene expression: Analysis of a dioxin-responsive enhancer system. *Proc. Natl. Acad. Sci. USA* **83**, 2802–2806.
44. Jones, P. B. C., Durrin, L. K., Fisher, J. M. & Whitlock, J. P., Jr (1986). Control of gene expression by 2,3,7,8-tetrachlorodibenzo-p-dioxin. *J. Biol. Chem.* **261**, 6647–6650.
45. Fujisawa-Sehara, A., Sogawa, K., Nishi, C. & Fujii-Kuriyama, Y. (1986). Regulatory DNA elements localized remotely upstream from the drug-metabolizing cytochrome P-450c gene. *Nucleic Acids Res.* **14**, 1465–1477.
46. Gonzalez, F. J. & Nebert, D. W. (1985). Autoregulation plus upstream positive and negative control regions associated with transcriptional activation of the mouse P1(450) gene. *Nucleic Acids Res.* **13**, 7269–7288.
47. Sogawa, K., Fujisawa-Sehara, A., Yamane, M. & Fujii-Kuriyama, Y. (1986). Location of regulatory elements responsible for drug induction in the rat cytochrome P-450c gene. *Proc. Natl. Acad. Sci. USA* **83**, 8044–8048.
48. Solt, D. B. & Farber, E. (1976). New principle for the analysis of chemical carcinogenesis. *Nature* **263**, 702–703.
49. Farber, E. (1984). Cellular biochemistry of the stepwise development of cancer with chemicals. *Cancer Res.* **44**, 5463–5474.
50. Astrom, A., DePierre, J. W. & Ericksson, L. C. (1983). Characterization of drug-metabolizing systems in hyperplastic nodules from the livers of rats receiving 2-acetylaminofluorene in their diet. *Carcinogenesis* **4**, 577–581.
51. Cameron, R., Sweeney, G. D., Jones, K., Lee, G. & Farber, E. (1976). A relative deficiency of cytochrome P-450 and aryl hydrocarbon [benzo(a)pyrene] hydroxylase in hyperplastic nodules induced by 2-acetylaminofluorene in rat liver. *Cancer Res.* **36**, 3888–3893.
52. Kitahara, A., Satoh, K., Nishimura, K., Ishikawa, T., Ruike, K., Sata, K., Tsuda, H. & Ito, N. (1984). Changes in molecular forms of rat hepatic glutathione S-transferase during chemical carcinogenesis. *Cancer Res.* **44**, 2698–2703.
53. Kitahara, A., Satoh, K. & Sato, K. (1983). Properties of the increased glutathione S-transferase A form in rat preneoplastic hepatic lesions induced by chemical carcinogens. *Biochem. Biophys. Res. Commun.* **112**, 20–28.
54. Levin, W., Lu, A. Y. H., Thomas, P. E., Ryan, D., Kizu, D. E. & Griffin, M. J. (1978). Identification of epoxide hydrolase as the preneoplastic antigen in rat liver hyperplastic nodules. *Proc. Natl. Acad. Sci. USA* **75**, 3240–3243.
55. Bock, K. W., Lilienblum, W., Pfeil, H. & Eriksson, L. C. (1982). Increased uridine diphosphate-glucuronyl transferase activity in preneoplastic liver nodules and mouse hepatomas. *Cancer Res.* **42**, 3747–3752.
56. Shor, N. A., Ogawa, K., Lu, G. & Farber, E. (1978). The use of DT-diaphorase for the detection of foci of early neoplastic transformation in rat liver. *Cancer Lett.* **5**, 167–171.
57. Pickett, C. B., Williams, J. B., Lu, A. Y. H. & Cameron, R. G. (1984). Regulation of glutathione S-transferase and DT-diaphorase mRNAs in persistent hepatocyte nodules during chemical hepatocarcinogenesis. *Proc. Natl. Acad. Sci. USA* **81**, 5091–5095.
58. Williams, J. B., Lu, A. Y. H., Cameron, R. G. & Pickett, C. B. (1986). Rat liver DT-diaphorase: Construction of a DT-diaphorase cDNA clone and regulation

of DT-diaphorase mRNA by 3-methylcholanthrene and in persistent nodules induced by chemical carcinogens. *J. Biol. Chem.* **261**, 5524–5528.

59. Lapeyre, J.-N. & Becker, F. F. (1979). 5-Methylcytosine content of nuclear DNA during chemical hepatocarcinogenesis and in carcinomas which result. *Biochem. Biophys. Res. Commun.* **87**, 698–705.

60. Pfohl-Leszkowicz, A., Salas, C., Fuchs, R. P. P. & Dirheimer, G. (1981). Mechanism of inhibition of enzymatic deoxyribonucleic acid methylation by 2-(acetylamino)fluorene bound to deoxyribonucleic acid. *Biochem.* **20**, 3020–3024.

61. Cowan, K. H., Batist, G., Tulpule, A., Sinha, B. K. & Myers, C. E. (1986). Similar biochemical changes associated with multidrug resistance in human breast cancer cells and carcinogen-induced resistance to xenobiotics in rats. *Proc. Natl. Acad. Sci. USA* **83**, 9328–9332.

62. Manoharan, T. H., Burgess, J. A., Puchalski, R. B., Pickett, C. B. & Fahl, W. E. (1987). Promoter-glutathione S-transferase Ya cDNA hybrid genes: Transcription and associated enzyme expression in mammalian cells. *J. Biol. Chem.* **262**, 3739–3745.

63. Scully, N. C. & Mantle, T. J. (1981). Tissue distribution and subunit structures of the multiple forms of glutathione S-transferase in the rat. *Biochem. J.* **193**, 367–370.

64. Guthenberg, C. & Mannervik, B. (1979). Purification of glutathione S-transferases from rat lung by affinity chromatography. Evidence for an enzyme form absent in the rat liver. *Biochem. Biophys. Res. Commun.* **86**, 1304–1310.

65. Tu, C.-P. D., Weiss, M. J., Li, N.-Q. & Reddy, C. C. (1983). Tissue-specific expression of the rat glutathione S-transferases. *J. Biol. Chem.* **258**, 4659–4662.

66. Li, N.-Q., Reddanna, P., Ghyagaraju, K., Reddy, C. C. & Tu, C.-P. D. (1986). Expression of glutathione S-transferases in rat brains. *J. Biol. Chem.* **261**, 7596–7599.

67. Abramovitz, M. & Listowsky, I. (1987). Expression of a unique Yb-glutathione S-transferase gene in rat brains. *J. Biol. Chem.* **262**, 7770–7773.

68. Hayes, J. D. & Mantle, T. J. (1986). Use of immunoblot techniques to discriminate between the glutathione S-transferase Yf, Yk, Ya, Yn/Yb and Yc subunits and their distribution in extrahepatic tissues. *Biochem. J.* **233**, 779–788.

# Subject Index

## A

α-Acetylenic DOPA, 48
γ-Acetylenic GABA, 41–42, 48
African sleeping sickness, 44, 71
Agmatinase in putrescine biosynthesis, 86
Alanine racemase inhibition, 54
Allopurinol, 66–67
4-Aminobutyric acid; 2-oxoglutarate amino transferase (GABA-T)
  catalytic role of subunits, 69
  concentration determination, 68
  localisation by autoradiography, 69
  mechanism-based inhibition, 40–41
  by γ-acetylenic GABA, 41–42, 48
  by β-alanine monofluoromethyl derivatives, 51
  by alternative substrate analogues, 51
  by dehydro-halogenomethyl GABA derivatives, 52–53
  by 5-fluoro-levulinate, 50
  by γ-fluoromethyl GABA, 54
  by gabaculine, 67, 69
  by monofluoromethyl GABA derivatives, 51
  by monofluoromethyl homo-GABA derivatives, 51
  by product analogues, 50
  by (S)-4-amino-4,5-dihydro-2-furan carboxylic acid, 67, 69
  by (S)-4-amino-4,5-dihydro-2-thiophene carboxylic acid, 67, 69
  by γ-vinyl GABA (vigabatrin), 42–43
Aminopropylcadavrine, 84
Arginase in polyamine biosynthesis, 86
Arginine decarboxylase
  inhibition by α-acetylenic agmatine analogue, 50
  in polyamine synthesis, 86
Aromatase inhibition, 68
Aromatic L-amino acid decarboxylase turnover measurement, 68

## B

Benzylamine oxidase, 62
Bovine plasma polyamine oxidase, 94–97
  inhibitors, 96
  prosthetic group, 95

## C

Cadaverine, 84, 95, 99
Caldopentamine, 84
Canavalmine, 84
Catecholamine biosynthesis inhibition, 58
*Chromatium vinosum* nitrogenase regulation, 19–20
Citrate lyase, 15, 16
  regulation, 16, 17
Citrate lyase ligase, 16
  regulation by phosphorylation, 16–17
Clorgyline, 56, 57
*Clostridium sphenoides*
  citrate lyase ligase regulation, 16–17
  citrate lyase regulation, 16
  protein phosphorylation, 4
Copper-dependent amino oxidase inhibition, 61–62
γ-Cystathionase, 69
Cytotoxic drug resistance, 136, 138–139

## D

Deprenyl, 56
Diamine oxidase, 62, 99
1,3-Diaminopropane, 84
Difluoromethyl DOPA, 45, 46
α-Difluoromethylornithine (DFMO), 44–45, 70
Dihydroxyacetone kinase phosphorylation, 15
*dnaK* gene product phosphorylation, 4, 19
Dopamine-β-hydroxylase inhibition, 62

144

## E

(E)-β-fluoromethylene-*meta*-tyrosine
    (FMMT), 58
Enzyme-activated/mechanism-based
    inhibitors
  catalytic mechanism studies, 69–70
  copper-dependent amine-oxidases,
    61–62
  definition, 30
  design, 39–40, 41–42, 43–44, 47–48
    alternative substrate analogues
      and, 51
    enzyme products and, 40, 48–50
    microscopic reversibility principle,
      40, 48–50
    positioning of latent reactive
      groups, 52, 53
    substrate analogues and, 40, 43–44
  dopamine-β-hydroxylase, 62
  dual enzyme-activated, 58
  duration of action, 36
  enzyme active site, peptide residue
    identification, 69
  enzyme localization by
    autoradiography, 69
  enzyme turnover measurement, 68
  enzyme-inhibitor adduct
    identification, 35, 38
  enzyme-mediated turnover, 35
  flavin-dependent enzymes, 54–55
  glutamic acid decarboxylase, 47
  histidine decarboxylase, 46
  β-hydroxydecanoyl thioester, 36–38
  kinetics of inactivation, 31
  lag time and, 33
  L-aromatic amino acid decarboxylase,
    45–46
  microinjection into specific neuronal
    tracts, 70
  monoamine oxidase, 55–61
  neuurotransmitter turnover
    measurement, 70
  nomenclature, 30
  nucleophile addition and, 33
  ornithine decarboxylase, 44–45
  partition ratio, 34
  PLP-dependent decarboxylases,
    substate analogues, 43–48
  PLP-dependent transaminases, 40–43

  polyamine oxidase, 61
  in polyamine research, 70–71
  protection by substrate, 33
  saturation kinetics, 31–32
  specificity, 35–36, 47
  stoichiometry, 34–35
  therapeutic uses, 71–72
  thymidylate synthase, 63, 66
  xanthine oxidase, 66–67
Epilepsy, 43
*Escherichia coli*
  *dnaK* gene product phosphorylation,
    4, 19
  glutamine synthase regulation by
    $NR_{II}$, 18–19
  glyoxylate bypass operon, 7
  isocitrate dehydrogenase, 5–6
    kinase/phosphatase, 7
    phosphorylation in intact cells,
      9–10
  isocitrate lyase, 5, 20
  malate synthase, 5

  protein phosphorylation, 3–4
    as phosphoserine, 3
    as phosphothreoninbe, 3, 19
    as phosphotyrosine, 3

## F

Flavin-dependent enzyme inhibition,
    54–55
5-Fluoro-levulinate, 50
5-Fluorouracil, 63, 66

## G

Gabaculine, 67
GABA-transaminase turnover
    measurement, 68
Glutamic acid decarboxylase (GAD)
  concentration determination/
    mapping, 69
  mechanism-based inhibition, 42, 47,
    48
    by acetylenic GABA analogues, 50
    by α-methyl-*trans*-dehydroglutamic
      acid, 47

# Cumulative Key Word Index

## VOLUMES 1 TO 23

(Volume numbers are shown in **bold** type. Page numbers refer to the first page of the relevant Essay.)

### A

ACTIN and MYOSIN, multigene families: expression during formation of skeletal muscle, **20**, 77

ADENOSINE, metabolism and hormonal role, **14**, 82

AFFINITY LABELLING, antibody combining site, **10**, 73

ALDOLASE, structure–function relationships, **8**, 149

ANAPLEROTIC SEQUENCES, role in metabolism, **2**, 1

ANTIBIOTICS, peptides produced by *Bacillus brevis*, **9**, 31

ANTIBODY, topology of combining site, **10**, 73

ARACHIDONIC ACID, alternative pathways of metabolism, **19**, 40

### B

*BACILLUS BREVIS*, peptide antibiotics produced by, **9**, 31

BILIRUBIN, degradation of haem and conjugated, **8**, 107

BIOELECTROCHEMISTRY, **21**, 119

BRAIN, metabolic adaptation in, **7**, 127

BROWN ADIPOSE TISSUE, biochemistry of an inefficient tissue, **20**, 110

### C

CARBOHYDRATES, of the mammalian cell surface, **11**, 1

### CARBON DIOXIDE, metabolic role in fixation, **1**, 1

CARCINOGENS, CHEMICAL, metabolism in the mammal, **10**, 105

CELL DIFFERENTIATION, nucleic acid synthesis and bearing on, **4**, 25

CELL MEMBRANES, turnover of phospholipids in animal, **2**, 69

CELL SURFACE
human leukaemic cells, **15**, 78
complex carbohydrates of the mammalian, **11**, 1

COLLAGEN (see also Procollagen) structure, **5**, 59

COMPLEMENT, activation and control, **22**, 27

CONTROL, of enzyme synthesis in animal cells, **13**, 39

CYTOCHROME P-450, hepatic, **17**, 85

### D

DEOXYRIBONUCLEIC ACID, repair, **13**, 71

*DICTYOSTELIUM DISCOIDIUM*, study of cellular differentiation, **7**, 87

DIFFERENTIATION
enzymic in mammalian tissues, **7**, 87
polysaccharides and lignin in plant cells, **5**, 89
studies of cellular, **7**, 87

### E

ELECTRON TRANSPORT, in *Escherichia coli* mutants, **9**, 1

151